The R.A.M.S. Library of Alchemy

Volume 37

The Transmutation of Base Metals Into Gold and Silver

David Beuther

R.A.M.S. Publishing Company

The Transmutation of Base Metals Into Gold and Silver

David Beuther

Produced by

Restorers of Alchemical Manuscripts Society
1987

R.A.M.S. Publishing Company

R.A.M.S. Publishing Company
117 Rutherford Lane
Stuarts Draft VA 24477

The Transmutation of Base Metals Into Gold and Silver

First Edition 2015

ISBN-13 **978-1511950381**
ISBN-10 **1511950382**

Image Processing by Philip N. Wheeler

Printed in the United States of America

Table of Contents

Dedicated to Hans W. Nintzel,

American Alchemist

and

Founder of the

Restorers of Alchemical Manuscripts Society

(R.A.M.S.)

Disclaimer

Liability: The publisher does not warrant or assume any legal liability or responsibility for the accuracy, completeness, or usefulness of any information, apparatus, product, or process disclosed. The publisher makes no representation as to the accuracy or completeness of the contents of this book and specifically disclaims any implied warranty of merchantability or fitness for a particular purpose. No warranty may be created or extended by written sales materials or sales representatives. You should obtain professional consultation where appropriate. The publisher shall not be liable for any loss of profit or other commercial or personal damages, including but not limited to special, incidental, consequential, or other damages.

The Transmutation of Base Metals Into Gold and Silver

David Beuther

David Beuther
Guardian of the Mint at Dresden and Expert in the
Practice of Alchemy for the Elector of Saxony

Universal and Detailed Account, in which: The
transmutation of base metals into gold and silver will be
explained clearly and precisely, along with an Appendix
containing unique alchemical copper-plate prints which
reveal the Art from beginning to end and a Preface which
documents Beuther's training and publications, along with
those of Dr Johannes Christopher Spregels of the Academy
of Hygienic Treatment and Medicine of Hamberg

Printed by Samuel Heyl of St Johns Church, Hamburg,
1718

Introduction

Philip N. Wheeler

This is a clearly written text that includes much alchemical information, including Beuther's experiment conducted at the Royal Laboratory.

As shown in the frontispiece, Beuther was Guardian of the Mint at Dresden and Expert in the Practice of Alchemy for the Elector of Saxony.

First printed in 1718, this work was selected for the R.A.M.S. Library in 1987 by Hans W. Nintzel.

Preface

by Dr. Sprogel

Highly Esteemed Reader:

Doubtless, it has been a cause of wonderment in many
places why persons expert in chemistry allow these
highly regarded writings to have been permitted to
be printed freely and openly, since the same ideas
are found to be highly regarded by the possessors of
a few manuscripts and here and there have been sold
for 1000 thalers or more, although copies of the
same are frequently available for sale.

However, this is precisely the reason why a candid
patriot has given the same to be printed. Moreover,
since it is seen that such a shameful practice has
been involved in the case of such manuscripts,
whereby many of them had no understanding at all of
the writing but nevertheless supposed that they must
have purchased something really valuable for such a
small amount of money, jeopardized all their earthly
possessions therefore and chased after pipe dreams,
so that, through love of the second best, in the
open publication of such manuscripts, they thought
to eliminate a small part of that which they could
not understand, by which means at least an attempt
was made to save as much of the cost of such
manuscripts, of which they understood how to be able
to obtain, from the wide sea of the earth,
graciously at every opportunity.

You might then see how to recognize our good
intentions and admonitions, which will not be
omitted where knowledge is found therein to return

the same to God and our neighbors, not forgetting
ourselves in the process.

What, accordingly, the author himself arrived at
(since he was born a German and had lived in the
16th century up to the year 1580 at the Court of the
Elector and Guardian of the Mint) indeed in the
chemical laboratory of Lowenstein, wherein he found
a complete history, which he considered to be
noteworthy and which he subsequently expanded
considerably.

I will point out that it is noted on page 565 in the
Records of the Court of the Elector of Saxony (under
the guidance of the Holy Church through Prince
August and Prince John Christian I, reported from
the year 1580 to the year 1591). Prince Augustus
authorized David Beuther as one adept in
experimental skill to subsequently act as an assayer
at the Mint of St Anneberg. In the Cloister,
attached to the same at present, where he intended
to have his quarters and laboratory, he roped off an
area near the wall, from which a little plaster had
fallen, and depended therein. Afterward he occupied
the place, although the plaster continued to fall
off, so that rectangular patches had to be removed.
All this has been described subsequently in Dec. 3,
the particulars of which are termed Pyrotechnics.
The first of these explains how two firings of types
of 'black sulfur' gave from a pound of iron in
ordinary water, finely pulverized copper after 28
hours of firing. Such copper was subsequently
dissolved in distilled water and in 4 weeks the
result was a precipitate of fine gold. Another
particular involved the transmutation of tin and
mercury into silver by using the same precipitation

process. The third particular comes by the Rule of Martial Antimony, which he made by means of a Crocus of Mars and Venus, prepared by casting gold and silver in the molten state. What he might have obtained by this process, nobody has ever been able to get out of him.

However, as he had prescribed in his work, he considered everything in order and set out in unhurried endeavor, limiting himself to a few (12) projects in one place or another, so familiar to him that he could oversee all of them. Accordingly, under the circumstance, all of these were added together in the end and he had completed many of them, and indeed with so little fanfare that the Elector himself had heard nothing about them.

However, Beuther no longer deemed his services to include the tests and counter tests, so long as his work in the case of the metals was completely unhurried, making use of the Elector's instruments, laboratory, and materials, at will. This was quite a heavy investment, particularly since nothing could be made subsequently of all that he showed them and, for this reason, his work was almost totally worthless, but nevertheless, when finally completed, the work was presented as such to the Elector.

Beuther was asked, when he appeared to answer the apparent complaints against him as a result of his 12 failures, about them; Beuther could no longer lie about such and had to admit plainly that the Elector had laid before him the stipulation and decision that Beuther should, in conformity with the contracts made with him, be held at fault, in accordance with his promise to disclose such information to him completely, so long as they should remain in Dresden and agree to give the Elector 10% of the gold and silver, as well as also to hand over the above for a certain amount of the metal, while the Elector would require half of the entire amount as a reward for his part. In the meantime, Beuther was held as a virtual hostage. This vexed Beuther so sorely that he decided to practice his profession no longer. When he revealed this decision to the others, things began to get much better, although he could conclude nothing. Because of this, the Elector was very displeased, although he had formerly been very kind to him; he was, however, allowed to leave prison and return to England.

However, a decision was reached to send him to Leipzig and he was taken to Gravamina. The decision was reached to first question him in detail about the process, threatening him with flogging for his misinformation and with cutting of 2 fingers for making false claims, and detaining him indefinitely so that he could not give this information to another monarch. This was read to him on a Saturday, the Elector having written these words to him by his own hand. Beuther had again given to me what came to me by God and Justice, even though I must undertake something with you on Monday which I would willingly reconsider. In the margin may be found these words: I beg you to prevent it from coming to this, and furthermore it be written I know very well that I can do it if you are present and I will also be able to do it even though you are not present.

All this information about Beuther, which this publication includes briefly, was supplied by the person who was at that time the private secretary to the Elector of Saxony and approximates what was the content of his report: My approach simply might be for the Elector of Saxony not to allow Beuther to vex himself, since he, as a headstrong man, might --- out of desperation --- cause a lot of trouble and considerable bother. In addition, Beuther might be persuaded to write a letter to the Elector of Saxony, in which he might bring his complaint and request the Elector to accept a solution in which he would no longer henceforth be required to remain silent. He would accept such a solution and again return to the House of Gold, as the Laboratory of the Elector of Saxony was sometimes called, based on its earlier reputation, although by then a protective cover, the Library, which was still in use at the time of Kunkel (the Year 1677) --- was associated with it. Since Kunkel sought to practice this art extensively, Beuther published the process in its entirety, even though it was quite different and affirmed the same conclusion.

Based on this process, Kunkel wrote a 4 page summary which Beuther had suggested in The Arsenic in place of a highly dubious work which the esteemed Elector had penned with his own hand. I must explain that I had no part in this nonsense.

Thereupon Kunkel continued the narration in the following manner:

Afterwards Beuther was reinstated in his former post and had been given the above mentioned guarantee of protection, and he wrote a letter containing particular data to the Elector, begging him to help him by sending him 100 guilders and pay within 8

weeks in fine silver or gold all of his past due
wages. Such gold as he had obtained from the Elector
had been repaid. In this way Beuther had also been
seen largely as the protector of the Art, even
though he was not completely responsible for
instructing the Elector until Beuther had finally
obtained a reasonable sum of several marks, which
are as good as gold, although he said (so the word
goes) that they were no better than horse manure.
"Now, I could help you with 9 pennies, which are
perfectly good", he said. Consequently, Beuther sent
away to his protector in order to get something
more, since afterwards he had to start a fire using
the bellows. His protection was perceived as having
come to an end, that Beuther had terminated his
employment and had, so to speak, jumped into the
fire. As he now returned according to his orders,
then Beuther lay on his back without understanding.
And whether, indeed, in all deliberate speed,
medical practitioners and priests were called in,
nevertheless nothing would have had any beneficial
effect on him and he would have died before their
very eyes. The arrangement had been concluded
hitherto and he had come to believe that he had made
a bad bargain. How Beuther had afterwards disposed
of his substance or how he concealed it, Herr Kunkel
could in no manner learn. Then the highly esteemed
Elector, Johann George, had said to him, and to the
others, that he had this tormenting matter, in which
the actions of a worthless fellow had also been
shown, which according to the Elector Augustus had
caused him to be burned in the venture. He had also
given orders to his secretary --- at that time, one
Lincker --- and after that to Kunkel. To burn in the
same way wheat was still left of Beuther's work, so
that his successors might not try to make use of

what they might find within. Which command, however, had not been carried out, since from the counsel of the principal director a part of the remembrances would be hidden and, hopefully, the original could again be found.

Herr Kunkel would have reported in detail, certain reasons, namely, to protect the highmindedness of the august Elector of Saxony who, in ordinary language, would have done him an injustice and caused Beuther to face an extended imprisonment, and moreover, would have made no contributions to the truth. The Elector of Saxony had shown Beuther so much kindness and, even more, might be so indebted to him (as noted in the above-mentioned arrangement that he had made with him and 12 other persons to observe those gentlemen who should not be blamed). In the first place, Beuther was a subject; on the other hand, the Elector had brought him up; in the third place, Beuther had begun the practice of this Art in the service of the Elector and had shown his loyalty to the Elector and had given that most esteemed gentleman his loyalty, above all others and had done so openly.

This far had Herr Kunkel gone in his description of Beuther's life and accomplishments. He prepared at this time certain extracts which he had drawn up from the work in his laboratory, recorded by the private secretary at that time, which can be found on display in the Laboratory Records on page 579 and following. And, finally, the account of Beuther's work is given on page 585, where he prepared the gold, and from all these accounts it is to be seen as also true that he has accomplished all these things. Whether, however, from the so-called black sulfur, as he termed it, he obtained a special tincture, cannot be ascertained with certainty. It must be considered how favorably inclined was the Elector toward Beuther, and how inconsiderate Beuther had been in return, and how he had placed himself in considerable mortal danger, although he was evidently right in this matter.

Herr Tuschky, the former grounds overseer for the Elector of Saxony, has also written in his outline of Gebaldis' Black Magic, printed in manuscript form, on page 89, about this Beuther: "He, David Beuther is actually a professor and has published a work called Particulars and another work called Lapidis (Precious Stones) and is a learned man, who has also produced for the Elector of Saxony much gold and silver, an Art, however, which obviously has not been completely revealed until recently, despite the fact that he knew very well that the Elector already had access to such a great artificer as he was. Moreover, the Elector himself has always offered him all kinds of good things, even specie (currency of the realm) since David Beuther was already pregnant with the idea, so that the Elector begged to sponsor him and after that, as was usual at that time, although only the sponsor should name the terms, although, however, the good words did not help at all, since Beuther excused his actions in the conviction, that, the same had been given credence much less employed, since many cabalistically questionable things had been allowed to creep in because of this. In the Year 1580 he was thereupon cast into prison because of practicing such Art, as already mentioned, where he had written in his own hand: "Imprisoned cats do not catch mice", by which it was meant to be understood that

he, since he had to remain in prison, could
accomplish no useful work for the House of Saxony.
Since he was now finally offered the opportunity to
transmute so much gold and silver as requested of
him from material available to him, he was also
allowed, without further restraint, to be able not
to take notice therewith that he had finally
resolved t take the desperate measure of drinking a
beaker full of wine containing strong poison,
against which no physician would advise, thereby
also terminating his life, so that the House of
Saxony would a the same time be deprived of both his
person and of his knowledge of an important Art".
Moreover, one finds in the refining fire of the
chemist a categorical judgment of him, which reads
as follows: Beuther is known as an adept of this Art
and is unable to lie to anyone. He has been to
Dresden and is the author of very much original
work, even though the Elector of Saxony frequently
threatens him with the hangman's noose, since he
intended to retain him as an examiner, while
directing him. He then took a large amount of the
tincture and gave it to him as a poison. Even in his
daily journal, which nobody referred to at all, but
the conclusion of which we hold valid, although the
author, as well as his Art (as explained from all
his judgments) fared well and prospered.

From all of these curious investigations, differing
to some extent on the circumstances of the author,
and on certain minerals, animals, and vegetables,
especially those which the sophists made use of in
preparing precious stones, which the author also
cited in the Foreword of Beuther's Testing and Art,
which he had also described on page 9 of the same:
"David Beuther, whose tincture was dyed a strikingly

bright cardinal color, even if revealing a considerable knowledge of this Art, would drive him straight to the hangman, though he might have given himself poison, which might have been more sensible".

His written reports, which might even now be allowed to be published, as noted in Particulars, might well then be the true natural philosophy and form of the Art and know how to maintain at a reasonable fair price. Then, so that the author might thereby prove himself to be a genuine authority and likewise show by introduction of the same how the Sun, Moon, Mercury and Saturn are extracted not by methods from the metals, but from their inactive state; and just as these materials are from the tree and the root, from which ordinary gold is obtained, many thousand times better than the gold and silver in nature and substances.

To be sure, he teaches almost the entire Art, since these excellent materials are also prepared by calcinations, solution, sublimation, purification, and coagulation, so that the Art acquires a special; degree of perfection and gives color to the impure unrefined metals, so that they will all again be in a basic solid form. Moreover, it is stated early in this text even more clearly, that thought shows that the entire Art depends on the availability of the proper material (not just one, nor even two, but all from a single root) and their solutions, so that nothing more will be necessary, except a fire and nitrogen, and that the same (even the materials) will be the gold of the philosophers, for which he states the truth from all the philosophers of the entire world.

In what is to follow in this report, as indeed in all reports on natural philosophy, it is the lack of knowledge about this process, which in fact does not pertain to a universal idea, but in particular might find a major use in the knowledge of the universal nature of this material, despite some abuse and misleading statements. When, however, attention is paid only where the philosophers' gold the philosophers' mercury, mercuric ores, the electro-minerals of Paracelsus, red cinnabar ore, and white arsenic (which of all of them alone, only the true material, and sulfur and mercury is that material, is separable as a salt) are concerned, it is implied from what was said that he indicated what he ascribed to the Art and to method of operation and how he showed the preparation and testing of the same, so that he would be able, when finally chosen for that purpose by God, to know immediately the proper and most practical material to use and to readily show that all of his processes were different, even though quite similar to one another. However, if one desired to demonstrate certain of his processes and also to include questionable and misleading historical processes, he should not then be considered an adept but should be considered even more of an imposter, since, in particular, the energetic Kunkel in his chemical laboratory had not even reported praiseworthy work of his own and the well-known Dr. Petraeus had not desired to report innovation in his newly published Preface to the Works of Basil Valentine, or that he had sold for gold some questionable processes now and then or that he himself had seen some of the same works in manuscript form, the author who had communicated them to him having originated them. For the two processes involving Mercury and Jupiter, I have given this David Beuther 800 German dollars of pure gold, so that I can now demonstrate, using his hand-written notes, that he assured me are faithfully transcribed, that I have found it as he also said and not otherwise, in the year 1605. Thus, it is primarily from the questionable and misleading historical processes that particular processes are not to be considered the same, although he, too, like that great King of the Arabs, Geber, in his report recorded before the forming of the Art, although nothing, however, before the others. Moreover, the principal preparation of one or another of the ingredients from the universal material had already been carried out, as had become clearly evident from the above-mentioned work of Kunkel, page 580, which reads: "The white arsenic powder is no longer prepared for the Elector of Saxony and unfortunately, the whole Art rests upon it", as noted by Beuther himself on page 7. Where will particulars be found, when the universal is questionable? Thereupon he said that without the knowledge of the true universal material (I don't say the properties of lapidis, for I am of the opinion that somebody will still be able to learn to recognize the universal material and from it produce the particulars and still, nevertheless, not be able to make the largest gems in this manner) no effective particular type of gem can be made. He who knows how to make this same white material can make his process successful, while others cannot. For that reason, therefore, his process made him no imposter, since he indeed allowed no one to copy it or print it who did not verbally disclose the Key and who entrusted no one to elaborate on it, even those who understood the spirit of the idea, as

conceived above. Furthermore, the report of the
energetic Kunkel can be shown to be correct in all
its parts, since it clarified, from many passages
found therein, that probably do not adequately
reflect properly on various aspects, or even on
enemies which he would perhaps have liked to have
discussed --- men such as Oerthlin, Schirmen, Jacob
Beuther, Hans Weinhold, and Heidler, to whom Kunkel
referred on pages 654, 572, and 585, and might well
have proven, or perhaps only suggested, that
patently false material provided no strong evidence
at all. Besides, Kunkel's report clearly states in
many places that he was guilty of no deceptions, but
had been a true adept, who had made much gold and
silver, although the method if such Art would not be
disclosed at this time, even though unreliable
processes continued to be reported. Finally, the
report of Dr Perraeri had been obtained and found to
disclose the time, whether he had not verbally given
at the same time both the Key to his process and the
purchaser and whether he had not found all this
information to be correct, or whether this in itself
had been capable, even when he, too, had the Key, of
disclosing the process according to the proper state
of the Art.

As, however, the various circumstances became known,
and a few would become known, that he had found in a
well in Dresden of this Art, he came forward with
the Tincture --- though others had described it. He
had obtained the tincture as a strikingly bright
cardinal color and found it to be a poison. On
account of defects in the otherwise satisfactory
report, it is difficult to separate. However, I do
not believe that a philosopher is particularly
likely to think this discussion appropriate.

Also it would not be necessary for us to judge his performance and why he kept his Art so secret and risked death himself rather than clarify the matter or disclose it intelligently. For who has seen inside God's judgments --- or who has ever been His advisor.

As a postscript, this written report is offered as proof of the strange alchemical behavior of copper, which was discovered by the same Beuther and also warrants authorization of his manuscripts, although in his writings he does not refer to it directly. These are the same, I say out of singular curiosity, which demonstrate completely and in great detail the entire philosophical work and actual material, which can only be understood and depicted clearly and taught as an entire process. I should also offer at this time a clear explanation of the same, but I am afraid that the whole matter has become more muddled than clarified. And I cannot in any way understand how, even only briefly, an even greater man has done it, by being able to work with incomparable chemical copper and such sophisticated additions, sublimations, and imbibitions, which are actually blamed on a natural philosophy that has never been considered. While it must be a general rule in this work, as stated by Arnold de Villa Nova, and again on page 66, that only a few minerals, a few medicines, an arrangement, a work, and some equipment, along with white arsenic and burning sulfur, which were to be made at the same time, as Geber had said: "There is only one mineral, one medicine, one digestion; and in this our entire work consists, to which we add nothing unfamiliar, or take nothing away, without removing excess impurities therefrom in the process".

Furthermore, nothing remains for us to do but to mention once more, to the reader to do his best to follow the processes of the author so that they may be coordinated with true philosophical matters. So long as, and until the same are confirmed for the first time, and afterwards at all times, to be correct, from which Dr Helvetius, in his Vitulo Arco, in Mutatis Mutandis, says: "I beg you, my friends, that you recognize the actual real materials, and do not forget them and their rule in the investigation of this Art, and give attention to their irreversible change while in contact with fire, for you will never find them without this information, live well". Reported from my study room this St John's Day in these old lands, as Steiner has written, thus the true solution will be found.

The Year is 1718.

Dr. Sprogel.

Special Arrangement Pertaining to Alchemical Copper

Plate No. 1, has no special relevance.

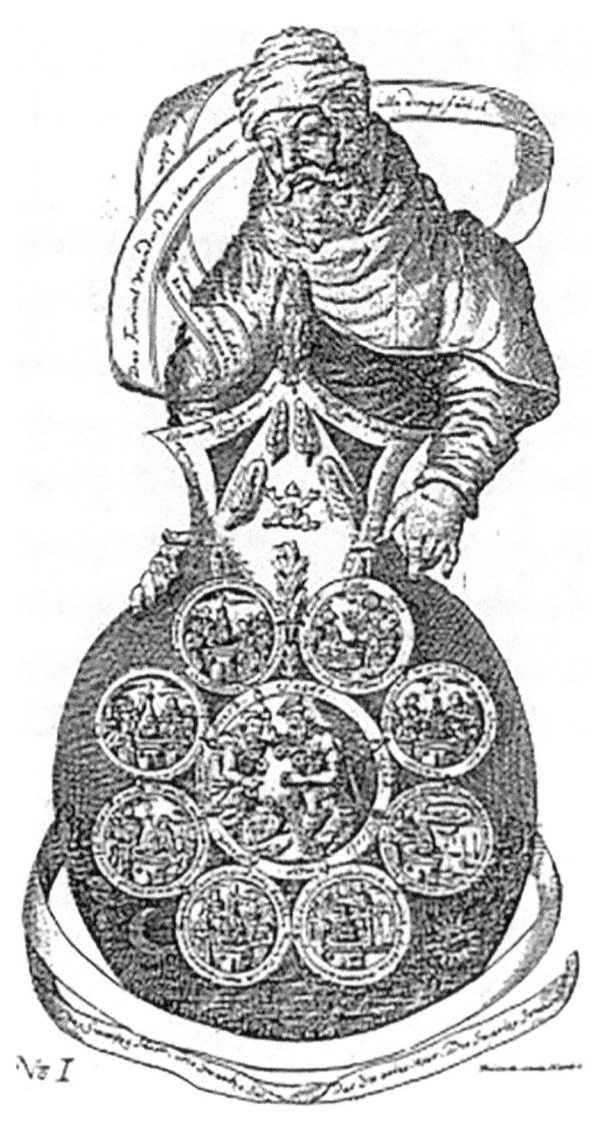

Plate No. 2, (a) Here is the final word on white minerals and the initial word on red minerals. The earth remains. The water washes. The fire purifies. The spirits go in;

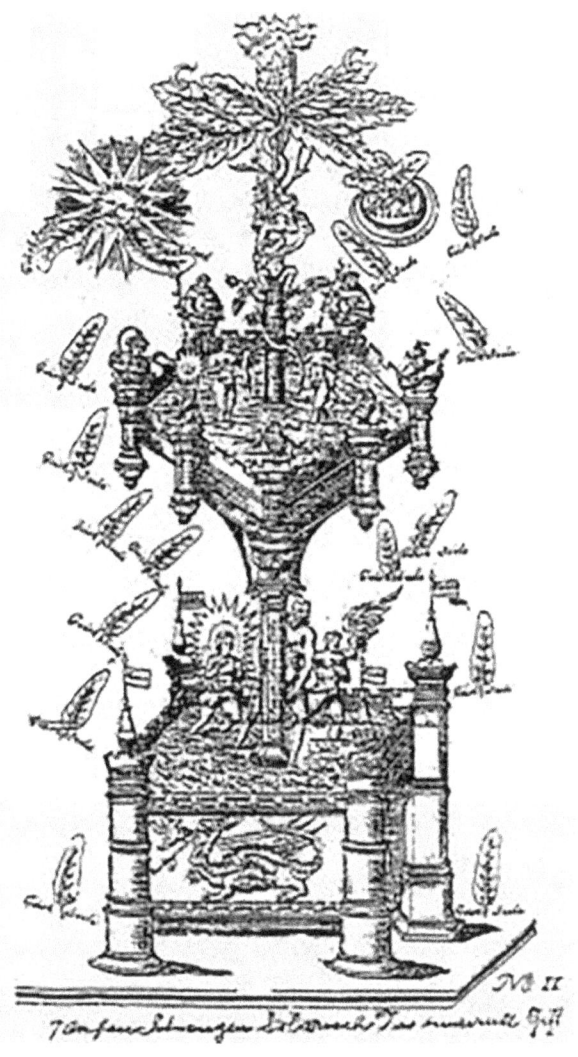

(b) Be on guard against the mouth of the hot-tempered. Avoid the presence of rascals and hot-heads;

(c) From the sun and the light, and the red gum, which is so brilliant; also in like manner from the moon, with white gum added to it, and the manner of natural brimstone, and I mention all of this sincerely, and I also name Scybric [Kybric] and Ryben and many others, and it is from these that I prepare my tincture;

(d) At the bottom is a heated place, as well as a snake in a fountain. Its tail is long and it has broad wings. It is ready to fly in all directions. Beware of such a fountain all the way round it. For your snake might get out of there. In case it does get out you will have lost the art of working with minerals, which is your motive. You must know that. Also what is in your furnace is so perfectly clear. And what is your serpent, with such a long tail? Although such work is of little use, your fountain must be able to burn in clear water. See well to this costly Fire, for your fire shall be able to burn in water. And the Water shall be washed with fire. Your soil shall be poured over the fire and water shall be accompanied by air;

(e) Now all this shall end in decay and bring a snake for deliverance. In the first place, black as a crow, he shall lie down in the bottom pit of Hell. I swell up like a toad, which lied in the mud and bursts open when blown up too much, which dies as though bitten by a snake and passes through many shades of color changes and finally becomes white as a bone. From the water contained therein, make him pure and free from his sins, and let him drink less and less, for that makes him beautiful and white, a whiteness that is a consuming fire. For here one sees the whole sequence from start to finish, from white gems to red gems, for here I have established for you the proper basis for this.

Plate No. 3, (a) Take your Father Pheobus, so shining and bright, who sits so high in Majesty,

with his rays shining down on you so brightly, in all places where he is able to reach, for he is in truth the father of all things, the preserver of all life --- of plants and roots --- and makes all nature flourish with activity, and he is also the balm for all wounds. To obtain this precious gift, one has to be very careful, for I say to the learned and the wise, that a genuine soul is his nature, which God Himself has shaped by His own Hand, and Magnesia is his Bride, this you will come to know for certain. Now I will here begin to teach you a way which will not long remain, the which you will seldom desire to take, yet mark it well, all this I am saying to you;

(b) Tender Phoebus, in many places with his rays, so clear and bright, uniting all this by his very nature, which is the source and mirror of all kinds of light. This Phoebus has many attributes of his nature which are very difficult to understand and where you cannot receive any understanding of what is right, do not attempt to find the Philosophers' Stone because you formerly craved it so much, I advise you. So mark well all that I might say to you and make thin that which is thick, and then it will be very well with you. You must understand what I mean and have very good self-esteem in this matter, for if you let your work count for very little, this will make you very sad. As I have said in this manual, he has very many natures of which I am aware, some recently discovered, others long since known. Heed his advice, as the philosophers do.

(c) In the sea, in all seriousness, dwells a little bird called Hermetis (Hermes Trismegistus), which can spread its wings in 4 colors, and can be found there, steadfast and durable, even though all its feathers are gone, for its stays as still as a stone;

(d) Here there is nothing but white and red, and the stone becomes alive which was dead, all of them and every one, in truth, bending only with difficulty, yet soft and malleable, and is now well understood, and his countenance is reflected upon.

Plate No. 4, has no special significance.

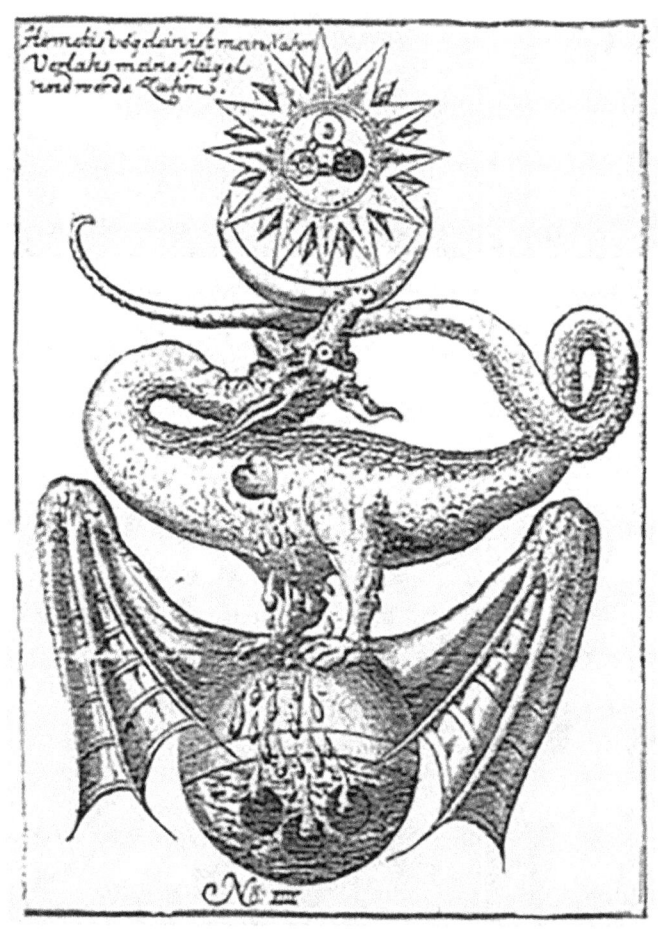

Plate No. 5: I will tell you, without telling you anything that is not true, about my behavior, since genuineness is my father and magnesia is my mother, nitrogen is actually my sister and copper, in truth, is my brother. The serpent of Arabia is my example, which is indeed the leader of all this, which I formerly held to be stubborn and wild, but now know to be tame and mild; the Sun and the Moon with all their might chastise me, who though myself so brilliant. Flights of fancy so led me hither and yon, until I thought that, with all their mightiness, they brought me down to the place where they wanted me to be. The blood on my hands is white, causing now both pleasure and happiness, and loosening me from the weight of stone which held me down, even before I was ready to begin. Now make hard that which had been made soft, for fire had surrounded it. From my blood and water, as I well know, comes all the fullness (abundance) of the entire world, which then flows to all sorts of places where it finds it has the greatest favor. It flows everywhere in the entire world and spins around like a ball; but you must understand this most of all, so that you will be aware that your efforts will not fail. Because of that you must learn before you begin, what it is --- and for all of your generation it has a great many interpretations, but is, however, of a single nature, which you must regard as three separate entities combined in a Trinity and bring them all into one single entity by the use of the Philosophers' Stone, which you see here and now.

David Beuther's

Philosophical and Chemical Writings

To God Alone the Glory. In the first place, from the
start I will bear down on each of you so that you
will consider that all wisdom comes from God and
this prize towers above all earthly endeavors.
Moreover, from God will be available such things,
and nothing will be too great and nothing will be
charged against him as error, so that then from the
first it lies within God's Omnipotence, for it is
stated, as in the case of a farmer, who plants the
seeds and knows the Tincture Process, for sure, so
that the Lord is not lacking in power to prevent bad
weather on render floods ineffective, especially by
such high favor, which the greatest treasure of the
world will also have, in particular, loyal people,
earnest prayer and perception of Nature. I will now
herewith write in this work about my attempt to
bring the fundamentals to you, and also thereby to
teach you how true are the presumed particulars of
the processes of the alchemists, which are the same
that one sees daily in keeping his promises. It is
the knowledge of their power and the fabricated
falseness which I will first of all attack --- that
most sacred Fundamental of the Sophists will I
attack by resorting to the true basis of nature and
prove through the evidence of God's Holy Word as
found in the Holy Scriptures themselves, from which
a righteous person can form a very good conclusion
that other lesser, though not completely useless,
persons must, nevertheless, not know the material on
which alone stands this Art, without which no true
knowledge can be found and, much less, shall have

the knowledge of the actual preparations. And finally, these fundamentals constitute the highest tests of Mercury, the Sun, the Moon, and Saturn, through which the tincture may be perfected, which are referred in the Philosophers' Book, which, however, are not correctly portrayed therein by those who do not entirely understand the philosophers, but only the vulgar manners of the literalist and consequently their pretended understanding might not be factual, so that what they suppose to be true Art would appear to be the exact opposite. To refute this, I have now given the following comparison to their great folly and ignorance: Human being were initially created by God without sin, but subsequently were introduced to sin by the Devil, so that not only did human beings, who were made in God's image, lose their glory, but also both natures --- the godly and the earthly (human) --- were in this way corrupted, obscured, and shattered. Similarly, all metals are hostile to the nature of the Sun and the Moon in the first 3 states of both bodies because of their impure sulfurous, salty, and mercurial spirit, which I attribute here to the Devil, who entered into the human heart and stained its spotless nature, which was the copper, and mercury, losing thereby the luster and brilliance of gold. And now mention be lost not only by the filthiness of sin, but also their own natural medicine, that is, through their own flesh and blood, but they can be redeemed to become as the first man, even Adam, initially without sin. This medicine man was unto the entire human race, since all were sinners according to Adam's Fall, and for this reason there must be a new Adam to be the Great Physician, who must be much greater in virtue and power in life and soul, and His power must be greater in all endeavors than that of all men, so that His Word must be ordained by God, first by the indwellings of men by the Holy Spirit, entering into human nature, so that the Redeemer of Mankind deemed it necessary to be at one and the same time both a man and God himself, in order that through such a tincture the godlike image, that had been so lost through sin, would again be restored completely, as pointed out by John in his first Chapter: In the beginning was the Word and the Word was co-equal with God, and the Word was God. Christ was also the true material of the complete medicine and tincture, which through His Life, His Death, His Resurrection, and His Ascension into Heaven prepared the Way that He obtained the pace second only to God the Father and has become the tincture of all men, It is the same with metals, which through their own medical practitioners are cured of their leopard spots and become very pure and noble, and of better habits, and so most simple metals and their impure materials be then also subjected to refining by calcinations, solution, sublimation, putrefaction, and coagulation, so that they attain a higher state of perfection, while tinctures of the impure and diseased Humorem Metallorum again reach the Glory of the Sun, just as did the man Jesus Christ. The question now really becomes, even when I then take gold or silver, since there are not no better metals than these two, according to the sayings of the philosophers: That which in the Sun and Moon rests upon the total Arts makes of them one thing, of salt, sulfur, and mercury, which indeed are the correct materials, knowledge and art of which can thereby be made complete, as there is at present no other way, according to the philosopher Bucher, that it can be found, and this is the most fundamental principle of all chemists, so that their greatest miracles must have a common underlying idea.

To this I made answer that the philosophers had written the truth, but such words were, however, not correctly understood according to their intentions, since they still prescribe and say: Our Sun, Moon, and Mercury are not ordinary, but are philosophical I might add: Even when I add them to salt, sulfur, and mercury, I find that they are not ordinary, but in my opinion they are philosophical. Therefore, I say that it is possible that one can give the other without itself undergoing damage, that it may or may not have requirements when this phenomenon takes place, so that you do understand the philosopher Bucher correctly, so that all of the particular ideas are then found to be true. Where, however, this may not be the case, then your entire hypothesis is also worthless, which I will now demonstrate by showing the truth that such of your presumed understanding amounts to nothing, and your Sun, Moon, and Mercury are worth nothing here.

We know that as long as men live, and body, mind, and spirit are in harmony with one another, then that man is a fully complete man. However, as soon as he dies his body becomes nothing more than a foul-smelling cadaver, which is no longer useful for anything, for his spirit and is soul (mind) have departed from him, in which all of the vital powers of this body reside, which keep him alive and maintain him through them. Now, however, life in him is no longer present, so that he also putrefies from his own very nature, and returns to the earth from which he came in the first place. Now shall he again live and be resurrected, and then his spirit and should must return to him as in the days of his youth, so that he might be a redeemed man and so remain forever.

We also on this point note that so long as the gold in his complete nature is not destroyed --- even as gold should be indestructible --- then it is firmly established, and unchanging when subjected to fire and freed from all contaminating elements, and we also see that gold will be melted away by fire, that is a mercury, runs and rolls out as another mercury alive, as soon as lower temperature causes solidification to set in, whereupon it cools rapidly and hardens. From the above it becomes at once apparent that the nature of gold is that a mercury and the ordinary mercury is entirely like the body. Now, however, the mercury is fixed in the fire, which although common, is, however, volatile and unpredictable, and does not produce mercury, but rather the spirit and tincture locked up in it, which, since it is sulfur and sal salis, from the ordinary mercury we are able to make through sublimation with sulfur, and stratifying with sol or luna this self-same sublimate and thus the mercurial extracts, through the help of the sulfur, the spirit or tincture solis by itself, and becomes Sol or Luna, after which it will become stratified with one of them; on the other hand, however, the Sol or Luna will become Mercury, which is volatile and unstable in the fire, so it can be smelted, like all those others, and thus the sublimate is prepared by my methods that are known to allow a profit for the work. From this it now is kept constantly that the body of the solis, that is the Mercury, the ordinary mercury, is entirely like the body, and of such shape, that it is worth nothing to the Art, even though a few tinctures may be made therefrom, but since sulfur and Sal Solis are used, this give to the Mercury the understanding of the meaning of purification by fire.

As soon as we now make the Sol into Mercury alive, then we have the best quality and tincture that we could desire, there in the twinkling of an eye, and the Mercury alive becomes unstable ad volatile, and extracted from its tincture and divested of the same; moreover, from a good thing, a bad one will come --- to draw a comparison, dead men are without vigor and might and have nothing to be able to give to another. Or the case of one having one of his own arms detached and is given one from another person. He can still not use it to help that other person. One might, however, also speak of the time when I take afterwards another Sol and restore his spirit to him and, by means of this Mercurium Solis extract it and coagulate it together, and as the philosopher speaks about it: Sulfur coagulates Mercury. Even according to your presumed understanding, you must nevertheless again take a Sol and then insert his spirit into the Mercury. Regardless of which use you desire to work, however, he would then bury the last, will not direct one blind man to another, and my words, which have nothing to add and which can give nothing to anyone else, will become true. Therefore, your fundamental concept is repudiated and you make a Mercury from the Sol and extract your tincture from it, although you will be considered cheap, as you would add the Sol to your tincture.

Examine now your work to see whether you can make something noble from something that is vulgar, or an angel from a devil, or from the first Adam, Our Lord and Redeemer, who is also invariable and steady like the Sol, redeeming the fallen Adam. As then your fundamental concept consists of the Mercurium Solis, even as a pelt from your ermine. Where will your particular concepts remain, when the universal idea proves to be false? What farmer is such a fool as to dry his seeds in the fire before planting them or then soaking them again in water and then planting them and then expecting them to bear fruit? Or let his overseer dry them and then plant them, or graft them onto dry stems, so that the sap dries up?

Moreover, who is now such a fool as to make a living tincture from a dead metal? Has a dead person at any time ever raised up a living man or can one sinful man redeem another? For Jesus Christ alone, who was without sin, was born of a virgin without the assistance of a man, and there in a lower form of man was hidden the supreme power of God, which however might not be displayed to help men to become holy before his suffering, death, and resurrection took place, became through the other degree of perfection for the everlasting tincture, through which we all could obtain the blessing of eternal life after partaking of this tincture.

Moreover, we must from the philosopher Bucher understand this and so must search for another Sol, from the first all-powerful tincture, without fault or defect, born without sin, its metallic character indestructible, which had known no sin, nor still shown any lack of perfection, but is still in its initial disclosure, and in itself contains the Heavenly Balsam, the same as that expressed in the principal asters. Such material must be recognized from the very first, for without such knowledge it is not possible for anyone to come to everlasting life by means of this Art or to establish that which is true and everlasting.

Now the material must be covered over by a fluid (which philosophers choose to call Acetum, Aquafort, Quintam Essential Vini, and a thousand other names, although it is really only a single substance) and be prepared according to philosophers' directions by suffering, crucifying, and resurrection, without putrefying with a fixed body in his Art, just as the dead body of Christ underwent no putrefaction, although it was dead, and came into existence as our tincture.

Moreover, in this matter, not according to the absurd understanding of the ignorant, who desired to extract the spirit from the Sol, and from it prepare Mercury, which is corruptible, and no longer capable of resuscitation, to say nothing of what their particular concepts should being out, as long as they were working in invalid things, and also in Saturn, Jupiter, Mars, Venus, Mercury, alumina, saltpeter, antimony, salt, cinnabar, sal ammoniac, sulfur, plants, minerals, animals, and numerous other things of the same type, which, however, are all completely invalid and objectionable, and in which no truth of the Art can ever be found, the same in aceto, aqua fort, or spiritus vini, or as it may be called, according to the terminology of Arnold de Villanova: When however, we take Sol or Luna, at the beginning of our work, but we will never find a water which of itself alone will work, but we must take a material, which of itself alone will work, but we must take a material, which is much better than Sol or Luna in nature and

41

properties, like the person of Christ, to reckon for us, for at all times that person shall make whole a number of others, who must be nobler and better, and then all will be made whole. As long, then, as gold itself is; for the most part, deprived of its tincture, and the first Adam must still be the same as anyone else, as we also know our own Adam, although in a thousand parts not so powerful as his first material of tincture, as we are or become men not unlike the first Adam prior to his fall, since likewise on account of sin the ordinary Sol cannot be a tincture. It was then, before forming a tincture, itself tinctured by its first tincture, the nature of which lies in this thousand virtues and is to be counted against the ordinary Sol.

The same material or Sol is needed in order to begin our work. We must search for it and take it from trees and roots, from which Sol is obtained, to which all philosophers direct us, above all others, Paracelsus, who is referred to in the Book of Minerals. When alchemists wish to find mercury, sulfur, and sal salis, as they then wanted to find the tree of gold and its roots, where they might be obtained chiefly, they should know that it is the proper Mercury, sulfur, and sal solis which generates the Sol at that point. In short I presume herewith, through sufficient fundamentals and preservation of the Word of God, and through additional understandings of false alchemists, together with their unreliable processes and anticipated results from their entirely inaccurate fundamental concepts, to be sufficiently answered. From this it then follows that their other numerous assumptions are not valid, since their presuppositions are not warranted. And, whether or not, from time to time something of value was found in their processes, it is nevertheless then according to the writings of Paracelsus that these interpretations and opinions constitute more of a hindrance than a fathering and even when they are able to further advance such a cause, then they can never arrive at the origin of it. They have no correct information and are able to say nothing about from whence it comes, for when they are asked how such a thing could come about, or even how they know that it has, or that such is indeed their process, then they are at once taken aback and can no longer guess if it is true or not, I cannot say with certainty how their answer comes about, for indeed I do not know. I have also found it described and indeed have labored over the problem of being able further to indicate approval, or even their rationale, according to nature, since it might be regarded as cheap, for often they were the objects most labored over in the cinnabar mountain (vitriol, antimony, sulfur, and similar metallic objects) which contain within themselves many times solar spirits, which are formed during the birth of such mineral in the veins of the earth, in which also the Primum Ens Solis (initial appearance of the Sun) is not wholly completed and continuous, but often completes the same operations, although they no longer are of the same species. Thus, also, is their Art brought to an end, for then they are able to transmute nothing, and so they are entrapped again, and no longer are they able to transmute nothing, and no longer are they are able to arrive at a conclusion. As a result of such erroneous ideas, they can even lead themselves astray thereby, and seem always to be getting into trouble, until they are entirely ruined and die as a result thereof, for which the beautiful Art must take the responsibility unjustly, since it is unreasonable to expect them to bray about their own folly.

Such particulars, however, admonish me, just as do the grim Reaper and the deceased Bucher, for they were carried out by Sinners who are now dead, and through the same intercession will the State of Blissfullness be attained. To err is also of the head, just as the correctly prepared tincture, which is Jesus Christ. They bind themselves to the foul body of sins, but even as their faith is, so also will it be with the tincture, even though now and then, though such superstitions as theirs no doubt something advantageous will take place, or be helped, and even miracles can occur and one can learn, in truth, that the Devil can also sometimes perform miracles, so that it, nevertheless, comes to pass to the same men that they are deluded and are therewith in such an ungodly state that they are only able to wade into sinfulness that is even deeper and in the end they are completely undone and, to a certain extent, filled by the Devil.

Likewise, it also happens with such persons, who dare to attempt such important things that neither God nor Nature, nor even they themselves, are able to comprehend either the beginning or the end, about which they are also no longer able to obtain anything truthful and will also, along with the Art that they practice, come to no good end.

Summarizing, such things will have the effect of giving notice of the proper nature, that is, the actual character of a Sophist who, indeed, in reality, seeks nothing and finds nothing. To

conclude further, however, as one reads of those who
have acquired a rather true knowledge of the Art,
who diligently studies the First Chapter of Genesis,
where God spoke: Let the earth bring forth grass and
plants, which will reproduce themselves, and bear
fruit, and each bear fruit after its own kind, and
have its own seed after its own kind of earth. God
also spoke to all creatures, as the entire chapter
shows, and the context enumerates.

Who is now in such things a fool --- he who resists
the ordinances of God and masters Nature on he who
makes a great effort to do God's Will? He who would
like to witness lynching and restore vileness and
ugliness? He who would sow lead, tin, iron and
copper as seed crops? And plants gold and silver in
the ground. Paracelsus says: "That which a man sows,
that shall he also reap". If you pant gold, then you
will reap gold, but living seeds of gold are those
that must be sown just as God and Nature created
them and has given them to you before your very
eyes. We cannot create the seeds, but Nature has
already created them. We only plant the seeds, like
the farmer does, and thus from a single ear of corn
many hundreds or even a thousand ears can grow, and
we see the same phenomena in all of God's creatures,
that is, each brings forth its own kind and
increases in its own manner.

In conclusion, I wish to say to you that the entire
Art stands as a recognition of current materials and
their solution. He who knows both of these ideas
knows the secret of the entire Art. But for greater
facility, less trouble, and lower costs, seldom can
be found or understood, for the Art will then be
absolutely so easy, as philosophers say, that it is

a woman's work and children's play, and also we need nothing more than fire and nitrogen, and this is the material which is the same as the gold of the philosophers, as I have already previously described in sufficient detail, through natural preservation; what is, will be. The operations of this Art, however, involve calcinations, putrefaction, solution, distillation, coagulation, and preparation of the tincture, with the operations of coagulation and putrefaction covering a period of 40-42 days, before the blackness appears, and then solution in 70 days, whereupon it appears white. Coagulation and Fixation occur in 120 days, and thereupon the product becomes completely white, i.e., it is the finished tincture which after 40 days becomes quite stable to fire. At that point a redness appears, so that approximately 3 months and a few days are required for this work, which is then augmentation initially of 10 parts, the second of 100 parts, and the third of 1000 parts, and gradually by another 10 times greater through solution and coagulation with acrid water, according to the writing s of Aristotle: "Through solution and coagulation we increase its virtue. Solution is that which necessitates, the true power is not of concern here, while solution is at the center of the operation of our thesaurus". Moreover, one sees that the entire work involves nothing more than solution and coagulation with acrid water, while working in water containing Aqua Mercurii communis, Aceto, Quinta Essentia Vini, or the like, which are not at all suitable for other work. Then we are able to see and know that, since the metals increase, the nature of these things shows that they need nothing at all. Still there will be found those whose nature we must reasonably follow, as they will have it and make it necessary. Also dissolution, according to the writings of the philosophers: "Bring salt and metal together without corrosive materials, and without danger, and red and white fumes will then result". You should also say, "Nature of natures, take your course; nature of natures, undergo change; nature of natures, follow to a conclusion; nature of natures, adjust to the result".

We understand that the farmer requires nothing other than seeds and soil, and there must be adequate rain and fertilizer, like that which is equivalent to nature's supply. Consequently, he ploughs the field and sowed it and then allowed the sum and nature to take its course, which causes it to sprout without any further additions.

Also, moreover, when we have such Seminal Metals and their Seeds, then we seed them in it and allow the sun to mature them. And during that time we have intense heat, then the sun can give by spending itself for a year what the sun might not be able to do in many hundreds of years. It is also true that it follows that the farmer who sows no seeds will have no crops; therefore, if we sow no salt, alum, cinnabar, sulfur, lead, tin, iron, copper, gold, or silver, but the philosophic, living natural gold, with creates the same nature and presents before your very eyes, after going immediately to work and then following all the needs of nature, requires such instrumentalities as nature has in the soil, and as a result we do not need as many instruments as the fool might presume, but with an oven and a few instruments we can bring along the entire process from beginning to end. And no special work is required, as Count Bernard has reported that we

heat the bath for the King, who rules over all evil
men on earth, indeed even as a child of 6 years
might be able to do, even though it occurs only
once, but how and when no man is to be found sinful,
then the philosophers have sufficiently described
all the absolute necessities of Art of this type up
to the power of the fire, which becomes evident from
the color of the material in the fire, and must be
recognized. And how much of the Art, which is the
easiest thing on earth, even for children, so that
they may laugh on account of it and that old folks
seek it so foolishly, so that it is, nevertheless,
the most difficult thing on earth to be found and
also it is with such small costs to be achieved,
that the entire work, when we have understood that
for which we have sought, would not cost more than
three thalers.

David Beuther's

Appendix to the Exalted Art of Alchemy

First of all, while we know that under all the planets and stars, the Sun is the hottest and most radiant in degree, and also that all the planets receive their light from the Sun, which planets are also able to, by their brightness, illuminate the entire firmament in the upper and lower spheres, warming all things that formerly were cold, and then, through the influence of the mighty Sun, whose force and effect led to the birth of gold which we refer to as Our Sun. Then, just as the Sun operates in its realm, it also becomes our Sun here below in earth. Then God placed Fire on the Sun in the sky, by which the excellence and the power began to have their effect; moreover, God has given alchemists and men of power and might, which kindle the earthy Sun, which give the inferior planets and the metals their luster and brightness and which assert the mightiness of the spirit of man through the Word, as God Himself said: "Let us make man in a form which will be like us, and he shall have dominion over the fish of the sea, over the birds of the air, and over all the beasts of the field, and over the whole earth, and over all the reptiles, which crawl upon the earth". So men had now received the commandment from the Lord to rule, as well, over the metals, according to the Word, since God had said: "Have dominion over the whole earth and everything which is above it and below it". Also, men will, from the time thenceforth, have dominion over the Sun and its erring path, which can sustain the gold and bend others to its will and pleasure, as did Joshua by his commanding, since he had routed his enemy, that the Sun stands still, signifying that it is possible for men to accomplish all things, when they obtain from God the wisdom and understanding to do it, so he also received the same with assurance in everything which he wanted to do, just as happened to King Solomon, insofar as he believed it only necessary for the glory of God and for his next task. And that is also to be considered for the might of alchemists and to cut off the presumed argument of ignorant fools who say that metals will be propagated to men from God by transplantation, so that it is considered to be a patently false art, which none can practice or have knowledge on how to carry on. For the metals do not grow like other things, for God has created just so much of the same in the mountains as there should be, and no more; for this reason must it also be brought forth with great toil and difficulty. Oh, you stupid fool, to whom apply God's ordinances and omnipotent words, which in the beginning He spoke to all creatures, in which He said: "Be fruitful and multiply". These words God spoke not only to living animals and to crops, but to slimy creatures and beasts of prey, which you see very day in the coarse dust, which increase by self-propagation, and for this reason you believe it to be true. However, there are others, that you are able to come to know only by alchemy; then in this, put your hand to the flail, thresh out the grain, which you sow from year to year, and so you will remain an unskilled toiler with a crude flail, all the while having faith in nothing, holding only for certain your clenched fist and will have learned nothing more than what the flail tells you, being able to handle it only, or to grasp it with your coarse mind of a beast of burden, since you do not know anything else than hands to be put to work to clean out stables, or to put manure on the field, and you yourself don't even know the place on purpose why you should do that, or that the manure gives quality to the field, so that the seed sown in the earth can propagate and you are unable to understand that both the seed and the manure must undergo putrefaction and in this way enter into the composition of earthly bodies and remain these only as the Quinta Essentia, vel Virtus regenerative (the quintessential regenerative principle), the same principle which considers the earth to be its mother, in the same way that the straw eaten by animals becomes manure, while the seeds, which are sown also become food for animals, and the horses and the oxen eat it and, after digesting it in their stomachs, eliminate the waste material, and in this way the straw, sooner or later, is converted into manure. And during the time everything in the stomach of animals is undergoing digestion in the stomach of Mother Earth, so that all earthly nature is, and so remains, the Quinta Essentia, and the force behind all these changes and from all these many processes coming together in the earth, where every one of them takes pace, much goodness results and other things are invigorated and improved and as a result the earth becomes fruitful and productive, and the good seed can bring forth a good harvest, just as a wife when she becomes pregnant beings a new child into the world. When it then receives a new personality. Such cases, however, will be to you a matter of instinct. Thus you go about seeding the field, but even as a blind donkey can find the manger and knows very well who has produced the fodder for him, nevertheless, you want things which are much too high for you and which will destroy

you, and leave you full of doubt, so that you will
know neither the beginning nor the end, and it might
not even be possible for the experienced alchemist
to transmute metals, even though God has given you
the blessing of children to bear witness to this
fact which, due to your inability to appreciate
this, remain completely hidden from you, so that you
yourself do not understand what a great
accomplishment it is, nevertheless (even though man
is the greatest of all God's creatures) to be able
to transmute metals. Now, no children ever came into
the world, when you did not need an alchemist, but
you attached great importance to the wife (as the
receptacle of the sol in which the seed was
implanted by you and nourished to perfection for
you). Should you then not have the power to rule, as
God has decreed, that you should then not have the
power to rule, as God has decreed, that you should,
by ruling over a great many lesser things, to keep
all such things under your feet. But you blind ox,
you don't see very well, and you are going to be an
uncouth fool and back yourself into a hard place. It
is also in your stupid head to find so little
understanding that you cannot perceive the wonderful
works of God and for this reason there must also be
in your brain something that is impossible to do.
There are then metals upon which a growing nature
cannot act, although the Word of God is spoken as
well for them as for all creatures: "Be fruitful and
multiply", each of you, and it will remain proper
and well with you, as long as the world remains, as
also is the case with the Word of God, Have dominion
over everything that is on earth and remain unbroken
in spirit, for this is sufficient to undertake the
alchemist's Art and the possibility of also
transmuting metals through the use of God's Word,

which confirms it, and no one can be able to refute this statement.

If, however, you wish to end this discussion and think that such is not the case, then you will seldom reach a proper conclusion or will find that the deceivers will be so may who go around with false and illogical arts and count all alchemists as one to the benefit of the Devil, so do not expect to judge as a teacher, but rather as a fool, for it will be seen that not all cooks which carry the measuring vessels are also not alchemists, which spend all their energies in such endeavors, but the understanding of God and Nature, and they know what lies hidden in Limbo Terrae (for which see Paracelsus in Aurora, Chapter 16, of the Universal Materials of Philosophy), which can also come to the aid of Nature, and can also make good the defects of the same through the art about which others only dream, while they let themselves seem to be those, who carry mountains of gold in their heads and have each one of them found an old catatonic state where a slovenly person, known as a student in a laboratory, has made up and set in motion a process which proceeds initially as follows: Recipe, originally in code but now deciphered, following a pattern of usage, according to which children of the streets follow a fool, who goes ahead of them with a pipe and will suppose that it is a certain that by leaping up and down he has attained a good conception of the true art, and still it was held to be so well hidden that no one must see it and when they finally take it in hand then they will find themselves deceived, even by one another, and, yes, it is true, they are even able to dare to undertake for themselves Strohschneiger and Hudelmann's versions of the glorious work of alchemy, which is neither contrary to God or to Nature, nor to one's own knowledge; not to mention the fact that they should have knowledge of the production of metals, from where they originate, and how they came into being initially, holding such friends of alchemists as rascals and knaves, although an honest and candid alchemist, who is also a true philosopher, and knows how to test his fundamental knowledge of his art, first, by knowing heavenly things, thereafter, through the course of Nature, and, finally, by the thing in itself, as such a famous philosopher as St Paul was known to be has written in the 15th Chapter of the First Epistle to the Corinthians, who you must surely not consider to be among the deceivers.

If you will, however, believe a true alchemist to be a deceiver, then see whether his art which he practices compares with the Word of God, fruitful by its very nature, and, in the third place, the thing that he practices, in which he labors hard and long, then as long as the mouth is the informer of the heart, so long will your own words make the truth known, whether, he is a liar or not. He now says that he will make it from lead, tin, or a baser metal, or even from verdigris, vitriol, salt or other material of this sort, from which the supposed alchemist gives himself the most credit and even thinks only too freely that he is a liar who will deceive himself about his own accomplishments. For then the Word of God stands there before his very eyes, since God has spoken: "Each one produces his own kind". In this arrangement, all things and all advances in nature must remain within this framework, since what lies outside such a framework is the work of deception and lies. If now a Sol or

Luna wants to grow or undergo transmutation, then it must not be sown in filth or vitriol or lead, or other base metal, since God has spoken further: "Everything has seeds of its own kind, by which it can be propagated; It is very good that everything has its own peculiar seed". Thus, each thing in its own nature shall multiply and bear its own sort of fruit. Thus, in vitriol, lead, etc., so the art of gold is not in gold or nature, no seeds of gold or silver are to be found. Just as also an owl can never hatch a falcon, although they both are birds and, consequently, have some characteristics that they share in common, so, too, neither can a man be descended from a horse, although it often happens that unnatural births do take place by the interbreeding of two very different kinds of things, although God has, nonetheless, forbidden such in his Commandments, as set down in the 19th Chapter of Leviticus, that such things born in this manner should be burned up by fire, since they surely are more Sodomitic and diabolical than they are human on God-like. Therefore, they also bear the mark of the Devil, they are misshapen and deformed, and are unlike other animals. While ordinary men bear God's own image, these misshapen beings are not like that at all. Therefore, I say that you can recognize liars by the mixing of strange things. Then also the old ways say: Like rejoices with like, Nature delights in Nature. Nature embraces Nature, joining together things that are alike. And adding further:

> Others, not feminine, for they shall give you nonsense,
> Woman melts man, for woman and man are tinctured.
> It is the rule of fire, the fire of Mercury himself,

> By the tribe, twice as much, the Philosophers Stone
> is esteemed,
> In the tribe it is he whom they ask thousands of
> thousands,
> These three protected, then all volatile material
> caught
> Others dissolve, add together, and subdue.
> Spirits and all fermented materials of the world,
> And this destroyed the coagulated body itself,
> One, if to exist together, with only the body
> placed.
> Sun and Moon arise from the colors themselves,
> Male by no means generates women, if not
> maintained,
> Nor woman prepare men without them.
> If the Sun does not make it ready, if the Moon does
> not join in,
> If only put down, but arising thereupon because it
> chooses,
> Sun and Moon, seed of Mercury, and whether
> Seed with feminine aspects is joined together with
> mercury
> For they prepare pleasingly these two bodies born,
> Because origin and prayer will conquer the summit
> of everything
> If fetters if women with the Moon is the goal of
> realization.

Now then, that will be enough said about the office of Chemist and of Nature, and to confirm the possibility of transmutation of metals and how the metals each one, according to the Word of God, has its own seed in and of itself, by which it can propagate itself and bring forth its own kind, according to its own fruit, and in this way you can recognize a true alchemists and philosopher, who differentiates himself from an imposter and prevaricator by showing that according to God's Holy Word and Divine Plan and the course of the path He

chose when in his material body, if you are unable
to believe the evidence in a personal way, then let
the Word of God be sufficient fundamental truth for
you, and your presumed argument to the contrary,
Similarly, might it not also just be possible to
make metals likewise propagate? Moreover, when you
become worthy of such, the evidence will demonstrate
it. Since one should not cast pearls before swine,
as Christ has said, then believe also that, without
previous trouble, your unbelief and foolish opinions
will put no roasted pear into your mouth or have no
roasted dove fly into your mouth.

So I will now permit myself to digress a bit and
proceed further with my previous argument, in which
I told you that God spoke, as follows: "One of every
living thing shall reproduce its own kind, according
to its own nature, and be fruitful, and retain its
own seed in itself and bring forth according to its
own kind". So we write of the power of God's Word to
apply as well this power also to metals, especially
to the highest and best from among all of the same,
which indeed is our Sun. Shall it, however, not also
provide brightness and light to the least of the
planets and must it then be set on fire by the
alchemist's reagents on the Earth, even as they are
in heaven above by God Himself, who is indeed Lord
of the Heavens, but has given the earth to men, His
Children, who have made it over to please God, the
Father. Now, such an ignition cannot happen by any
such thing of itself alone, but through its own
nature, and the essential inflammation, no fire
could be found at that time, which could be the
power, the excellence, and the result of pure
activity alone, for it will help you to understand
the effects of the other planets, whether they also

are fiery, with their heat giving off nothing and taking on nothing, or like the Sun, or fire from God, so that the fire and the spirit are the same. For this reason also, as long as they are taken by one better than it itself is, then they can impart also to others their excellence and this will then be a light, which can then provide light to everything which is dark and obscure, because if their cleanness and ability to light up and make warm, and the other planets, which of themselves are the same as dead, can be kindled and brought to life and, therefore, on account of their supreme majesty, can be compared to God and Christ, which also would have given rise to the planets around the Sun, including the Planet Jupiter, and the Sign of Leo the Lion, and since, moreover, the Sun illuminates everything in the sky above with its brightness, and indeed produces the same result in the inner spheres, such as gold, minerals, rubies, plants, animals, lions, birds, and even men, and also Christ had the proper Sun in all its clarity, because of its brilliance and never-failing light which it brought and explained it and defined it, and was destroyed by inhalations of the Spirit of the Devil, but how Christ, both God and man, who was complete, so that we all can also be more complete, so that men and creatures have now the power, although the power of the Spirit, the Godhead far surpasses any power we have of our own. Moreover, the heavenly Sun, because of its great cleansing power, can being about more than all the stars put together, as also our earthly sun here below with us does to metals, and as Christ is the foremost among all men, and next to Him we are the noblest of all creatures, even better than the angels, for were are created in the likeness of Christ, who died to redeem the whole human race and has not suffered to let the will of that angel who was cast out, and also in earth and heaven, the Moon, and the Sun. To rank among the lesser plans, as with respect to Christ, and as (Luna Conjunx Solis) the wife towards the man, and by nature so completely united, and how very much more is the man than the woman, since it was the man who was created first and is thus Lord of the woman. In the same way the Sun is Lord of the Moon. (Woman was created from the man, so mercury is masculine, while sulfur is feminine). From this, then, we conclude that to complete our work, as to begin the same, of improving and shortening it, no other material should be taken than cinnabar, since the metals in vitreous earth are unsatisfactory and thus they must be treated with a better, higher, and nobler medicine, and the greater the infirmity, the greater the medicine must be, so that the healing process and recovery can also be more rapid and more complete. The entire operation may be briefly described in the following verse, first set down in the handwriting of Paracelsus:

There is a stone and still no stone,
In which stands the whole art alone,
It can be found in the ground and mountain
In which dwell ogres and dwarfs.
I tell you this of a truth and as a favor,
He who has this same thing has only a vapor.

And the splendor of the red lion,
Extracted mercury, completely pure and clean,
The same sulfur, I saw of a truth,
Had the rudiments of the art so completely.

Raymond Lully in His Vademecum (Guidebook):

The tincture was extracted along without Sol, which you can easily understand, so that you can read our book without annoyance, and can go and do your work more safely. Consequently, the following things are required for higher study; a better understanding, reliable information, hands ready to work, avoiding sophists, free and unhindered employment, sufficient ability, and means and a good philosophy. Bucher had them all.

Johannes Daustenius Anglic:

Cook the little man and woman together, until both become hard and dry. When they are not dry, different colors will not appear. Cause: the color will always be black as long as moisture is present.

Item Daustenius:

It is very fortunate that Nature will react according to well-established patterns, in which most substances are not completely destroyed by too much heat or lack of heat. Thus, when Nature reacts too vigorously or too little, but otherwise as I ought, then it destroys directly and that is not the same, consequently, as that which one would have wished.

In honor of the King and Queen, however, and indeed not burning them, for they did not flee therefrom on account of the intense heat. So, then, it is for you to be patient and long-suffering, which improves the Kind and his Queen in the government. Cook them until they become black, and afterwards white, and then red, and then this will become an actual medicine, which gives its tincture to gold.

Whereas our ore originally became dissolved the more it was heated and became a volatile liquid, after that the more it was heated, the thicker it became, turning into a very white powder. In the third place, the more it was heated the more colored it became, until it become a tincture with a deep red color and the entire operation was reduced to a process as simple as the extraction of water from an ore and then decanting the resulting liquid from the ore. There must, however, be a diligent watching, a continuous confidence. Thereupon, the whole thing will become self-explanatory and will also finally turn out to be to your liking.

David Beuther's
Philosophical and Chemical Writings
To God Alone the Glory.

Question: By what means should one dissolve the body of the Sun and the body of the Moon?

Answer: Through your own water or key.

Question: What is your key?

Answer: It is your own mercury.

Question: What is your mercury?

Answer: It is a metallic liquid which is an inner vapor of minerals, which is completely volatile so that it can also disappear into the air and diffuse into the air and after sublimation it is referred to

as the Water of the Discerning, you spring and
fountain, as well as your fire, for then your
Philosophers' Water is actually a fire.

Question: Where must one seek to find this Water?

Answer: In the mountains and in the ground and in a
large of minerals.

Question: In which minerals?

Answer: In those minerals in which the spirits of
metals reside.

Question: Are all of them good?

Answer: Yes, but all of them are different.

Question: Why is that?

Answer: Because? If they will al arise from a single
root, then, however, nothing can be extracted
without gold and silver; no tincture can be produced
from itself alone. They can all be converted into a
tincture of a medicine, but one of them, however, is
very much nobler and better than the others.

Question: Why is that?

Answer: The reason is that one of them is not equal
in quality to the others in the stars above and in
the celestial world in the embodiment of their
influence.

Question: Which of them, however, has the greatest
ability from God to act as a tincture?

Answer: There are indeed many spirits which under such circumstances do have the power to give metals a special color. These are the spirits of metals residing in a great many mineral ores.

Question: What are such minerals good for?

Answer: Indeed there are many uses. For one such metal has the spirit of the other six in it. But the most important spirits, however, are those of silver and gold, which live in base metals, such as cobalt, zinc, bismuth, antimony, and the various forms of garnet. The highest of all of them, then, are gold and silver ores and from them the noble water might come when they are found in an entirely natural and pure state.

Question: Which are the best and most serviceable?

Answer: The most dilute and volatile ores and minerals, since they lead to the spirits of the Sun and Moon.

Question: Do these spirits have more ability as a tincture than others do?

Answer: Yes, indeed. Certainly far more. Since they actually surpass the others in their ability to become fixed, and in their stability, and also they surpass them in volatility.

Question: In which minerals are ordinarily found the spirits of gold and silver in the greatest quantity?

Answer: As you have desired to know the origin and the source, then you will see that the spirit of

gold and silver is pure, unmixed with other living
spirits, and you will also see that silver is the
most serviceable for it is (even when unrefined)
pure, ductile, bright, volatile, and the lower the
silver content the better, since the red-gold ores
used are broken into gold and cobalt. Among others,
cobalt is often found, which leads to a little
silver, if indeed any whatsoever. Then after digging
in a mine, a white, soothing coagulated liquid or
liquor of the moon is found, which is formed
partially from actual cobalt and tends to increase
in amount and is inclined to be oily, like melted
fat, or even a little pasty, or is like a soft,
delicate, white painter's color. In this liquid,
however, lies our liquor, hidden, from the moon, for
us to labor over. When, however, such liquor appears
to give a mineral a red color, then this leads to
the Spirit of mars, even if, however, it becomes
green, like the Spirit of Venus. This spirit becomes
separated from its ore by spagyric methods, so that
it has within it the life of the earth, although
this spirit has also separated the pure from the
impure by the process of sublimation, or, in another
way, by the process of extraction, carried out with
rather vigorous extraction and the addition of
silver, leading to artificial flowers of lapidi and
tinctures resulting from their solution.
Question: How is this spirit driven away and
separated from its crude earthiness or terrestrial
nature?

Answer: In order that you can accurately separate
the pure from the impure and obtain each of them
separately, you must now take the liquorish water
containing the gold and silver and place it in an
efficient sublimation apparatus or an earthen vessel

where it is then sublimed in a sand bath, using a vessel made of iron and tin, and allowed to react slowly for several hours and then, according to the opportunity, the longer, the stronger, the spirit rises very gently through the neck of the still head or retort into the other globe glass and is even allowed to remain there, the metals being identified by their colors.

Then, take an ore and bring the spirit therefrom. You must tap it gently, so that the spirit can rise easily and go out of it. Or, take the mineral ores of gold and silver, crude or pure, in pieces or entire, carefully pulverized by rubbing on a stone, then hermetically sealed in a vial, melted, and then allowed to sit in a sand bath and made completely soft, compact and fixed, so that the gold becomes heavy and is covered over with a green skin, which looks like the leaf of a nut tree and is spoken of, as is the custom, as the Green Dragon of Hermes, or as the Green Corrosive (Omnivorous) Snake, or as the Green Lion, although Theophrastus called it the Tree Frog, imbibing an air of wisdom and coating his back with wax; but in reality, it is the White Lily, the water of the Wise, for from it comes the Elixir of Life, and the Fountain of Youth. When now this Liquor of the Sun is drawn out from the resin colored blood and the green tree frog is covered over with, making him so hot that his skin will burst, and from him, after he does indeed burst, comes our Lily, which grows even snow-white, like wool. Let it grow as long as it can. Take this with an instrument of gold, finely constructed, and when you have enough of the same, then continue the experiment in a spherical vessel which is formed like a round-bottomed flask and sealed, and its contents allowed to putrefy in our Athanor oven, made of brass, so that one can carry the experiment out in a dish or whatever he chooses. It is approximately one span and a half wide, well made of good iron or copper, like figures with two cupels, in which putrefaction takes place in a very mild heat, with very mild sweating, with low heat over a period of a month, resulting in a black color, then a white one. This is our Fountain and our Soil, in which is found that which you will harvest --- the Corpus Solis et Lunae (Body of the Sun and Moon), which can be further purified. After this stage of purification, more gold was added and the mixture was allowed to putrefy further. This is done as often and as extensively as desired. There is, consequently, a certain considerable amount of increasing the amount of ore. So milling takes place nine times, putrefaction occurring at each step. You must see the blackness, reminding you that you have to correctly solve the problem and must make certain that, when necessary, the gold must be added to the mercury. When all of this is now in solution, and then added to its own ferment, and heated for a long period of time with one another, until it is unchanging (that is to say, the gold or its ferment --- the gold or silver added as a leaven, so that its ferment is volatile and the volatile part, the mercury, is unchanging) and it was formerly considered a tincture, although not now.

The following involves three kinds of solar liquors which are very serviceable to the Philosophers' Fountain:

The first is called gold pyrites, which is entirely pure by itself and leads to no other foreign metallic spirit or any highly corrosive substance. The other is the crude and scaly form of hematite, which, however, must be pure. This is the most stable and durable Mineral of the Sun, whose greatest value is seen in a noble medicine comparable to the noble mineral, antimony, in which the Spirit of the Sun and moon may be readily dissolved to form a medicine.

In the third place, there is still the Mineral of the Sun, which is a liquor or water in very solid gold-cobalt, gold pyrites, or gold quartz, which is very mild and delicate in its first composition. It is found to have a red color and appears to be like a cinnabar ore. This is the Sugar of Theophrastus, which he mentions in his Thesaurus to act like gold, and with this mineral, the Sap of Gold is brought to the fountain and allowed to undergo change (as formerly pointed out with the Mineral of the Moon). This is now the threefold Mercury of the Philosophers, which is, however, its ferment, actually a very beautiful pure gold oxide, to which when a gold liquor was added and allowed to putrefy, a black material was obtained, so that the Green Dragon (which is simply mercury) which has been extracted from the mineral and blended with the Anatron Sol, which is actually the Lion of Gold, or gold oxide, which immediately turns it into the mercury and then into a double mercury. This is now the true mercury of the philosophers, which dissolves all substances if enough is added. When all this is finished, as all of the Philosophers have noted, the two waters unite to form one. In this white water, the substance dissolves to form the perfect solution, which becomes coagulated when acted upon by its first ferment until it, in its turn, becomes white, as well as non-flammable and indestructible, and even though it is actually fixed and stable as a White Tincture, it must, nevertheless, be further treated until it gives a red tint, at which point it is even better fixed and is then able to dye everything from white and can be looked upon further until finally, the deepest shades of red appear. Continuing this process and others, such as steadily heating over an open flame, produces no change, although a brightness is imparted to the flame, which first turns gold, then red, and finally a deeper red, but the substance itself is non-combustible, which is its most notable characteristic, since fire only serves to make it more stable.

Question: How many changes, then, can this Stone bring about?

Answer: Two. The first is putrefaction; the second is fermentation.

Question: Then how does fire cause it to increase?

Answer: The increase occurs in the metal.

Question: How can the gold cause it to become tinctured and make it better, since it was previously pure enough?

Answer: It makes it like a tincture and a lapidi, which thereafter is much more penetrating and its activity is increased so that 1 part per 100 parts of material can bring about the same amount of uniform coloring and can turn out more gold.

Question: Is there then no mineral on earth that this material cannot cause to be tinctured?

Answer: None. Of that which you have seen as the proper basis and, as I have previously suggested, all metals have originated from a single root stock and thus all can be brought to a single tincture of white and red (although the assistance of gold and silver are required). So you now know, and can imagine even, what kinds of metals are best suited, the most powerful and the mightiest, and you will soon discover the gold, silver, mercury, copper, iron, tin, and lead all have a single root or Mercurium. When occurring alone in a metal or mineral this ore is purer in mercury and is easier to bring out of its ore and also easier to purify than are the others. In this way, you will know, however, that mercury, while still in its ore, is also a very beautiful red color, resembling that of a ruby. Indeed, this mineral is also finely powdered and comes from its ore very readily, as mercury also comes from its gold ore, even more rapidly than other forms of mercury. The gold and silver in their original state can bring it out and make it similar to mercury. This is the only difference in the form of the minerals reported, that is, whether they all will have one kind of mercury and stem from a single root or a mercury before them will be sufficiently shaped solely by other characteristics. However, through the poisonous vapors in the pockets interspersed in the earth so that mercury from a mineral having another spirit might lead to a mercury of such a nature that it can be purified easier than the others and thus one sees in such a one this fact in the sublimation process, since the mercury is often in four different colors, all of

which are spirits driven thereinto. On account of
this, no mercury dissolves the metal. Then it will
be that false colors and impure spirits will be
separated and removed. The mercury, however, from
the living mercury ore is volatile and is even the
most volatile of all minerals and, also, without a
single thing, except a very gentle heat, not alone
for the purpose of driving it from its ore, but also
(by such a gentle heat) to readily purify and refine
it and, moreover, that it might equally have such
strength and other good qualities as the preceding
do not have. And when it now has also been made as
completely pure as a crystal as a result of
sublimation, then it dissolves much more readily
than other minerals, such as mercury, gold, and
silver, and combines with them, so that this is a
true saying, "Prepare mercury from mercury solution,
without ordinary mercury ores". But Philosophers
say, "Thus you see that mercury ore attacks the
metallic nature of all kinds of metals and fuses
them together and, for this reason, you can also
believe that it can do that to its mercury even more
readily". When the spirit of the pure gold is added
to such a rapidly heated mercury, then it makes the
gold like the mercury, too, and so one can proceed
with the other ores, as previously taught, to
prepare tinctures from other metallic ores, thereby
leading to mercury or the spirit of the gold. For
this reason, nothing is bound to any one certain
mineral, since from one such, as from the other, a
mercury is obtained. Then a convenient process will
result, which can then be utilized. It is also to be
noted that whether indeed a mineral containing
mercury taken from Nature or from another source
that can be repurified like the other, then we must
now take such a mineral, as Theophrastus taught,

namely the Electrum Minerale Immaturatum (which is such a noble metal) which by itself has no other foreign nature, then as long as the spirit of the metal will be of many kinds, many kinds of processes must also be required to refine the same. In pursuance of this, I will now consider the philosophical work to follow hereafter. In the Name of the Holy Trinity and to the Glory thereto be given. Amen.

Philosophical Work

Since the metals in the veins of the earth will be very impure, they must be treated by a very high and noble course of medicine, and then will be cured. The more persistent the illness, the more vigorous the medicine that will be required. Thus, for our Work, there is nothing better than can be taken than the labeled material and for that we must go to the Tree of Gold, as Theophrastus has shown us and has also stated in his book on the Minerals.

The old alchemists knew how to find sulfur and mercury from the Sun and how in the Tree of Gold and its roots he could indeed find it, and might even be able to obtain it cheaply, for the same, namely, salt, sulfur, and mercury, which is mercury, produce gold there. They also talked about a Tinctura Physicorum, in which the old Spagyrists before them had withheld controversial and tedious work, since they had obtained a much easier and better route where they had learned about the Astronomiae Concordance on unanimous opinion of the alchemists, in which Theophrastus speaks about the Lion: "Take now his ruby-colored blood and from the Eagle the whiten gluten, and after you have coagulated the two together, in this way you have obtained the Tincturea Physicorum". What the Red Lion was, however, he said --- at that time and in that same place --- only with obscure words, and there he described the Subjectum Materiae, saying in these words: "The material of the Tincture is a thing which you understand correctly from the Spagyrists, which come from one of two Arts taken from the Vulcani (the Shining One)", and might so still remain as I told you about in his correct terms, according to the old customs, and even that it is the Red Lion, named by many but known by few. Now, the same may be transmuted through the help of Nature and the skill of the artisan in works of the White Eagle, and also, that from a single one, two will result, over which the golden brilliance does not subsequently lighten the Spagyric over the two, viewed as one. If you do not know what the requirements of the ancient astronomers and the customs of the old cabalists (i.e., mystics) are, then you either report for further enlightenment through oral communication or else elect to look into the Arts of Vulcan through God's help or that of Nature. So much has been said about this, as Paracelsus would say that he used the short way of preparing the tincture from the materials from which all this should be made. It is so clearly described that those who do not understand it will not even be able to guess anything further and are unable to declare it clearly with their mouths. And although it is true that it is given to understand sufficiently the as wisdom, so it will nevertheless be imprudent, and even more foolish, to never learn from this during one's own lifetime. Thus, from the start, it was indicated clearly enough what the material is from which the tincture should be made and is stated in this way on the basis of one thing alone, and that is, that one needs nothing more than this (just as also the entire soul stands alone in Christ) and should say: "It is and is called the Red Lion". Where, indeed, we seek the same and shall find it, he says, in the Tree of Gold and its roots, since he leads us actually to the one mountain, which is nothing other than the actual Red Lion, and then to that which is no other metal or metallic spirit by itself than the cherished Spirit of Gold. And he, indeed, clearly and properly talks about the fact that in order to demonstrate these words, he said the following: "The Red Lion can be transmuted into the White Eagle, so that the one can become two, and moreover that gold with all its brilliance cannot lighten up the spagyric".

Note well, the word "Gold" and all its brilliance, of which he speaks and by which he means nothing else than the above-mentioned material, and not the ordinary melted gold, proving from this that, since he has directed us to go to the Tree of Gold and its roots, the metals increased, and also in his Archidoxa, as well as his little book on Renovation and Restoration, he said that the gold, indicated in the thousandth part, is not so powerful when melted by the furnace as in its ore, for as soon as this is done, the salt, sulfur and mercury are gone away. He also pointed this out in his manual. For this, we should take Erlectrum minerale immaturatum, that is, the most noble of metals. An item in the little book Electro-Minerals says that from the purer metals, nothing is retained of the art of alchemy. And in his Metamorphoses he also says the same thing about active gold and silver. Rodangionus, in his little book on Signs of the Zodiac, writes that the art of

Alchemy is not to be sought or found in metals which are of poor quality and inferior and, so also, will nothing be found in small minerals. Arnold of Villanova in his Alchemical Speculations points out that in no minerals of inferior quality and not even in ordinary gold and silver, and not even in other lesser things, can alchemical arts be found, excepting those in which the gold is the best possibility and satisfies the symbols of cinnabar, that is, it is in the Red Lion. Theophrastus also wrote about such materials in his Alchemists' Thesaurus, as follows: "We find in the veins of the earth a mineral which, in its first coagulation, appears to be red in color and in it lies sealed all the flowers and colors of the minerals, as well as the good qualities of all metals, and what might be the same in all of them, that is, in this single thing", and noted as such in the same sort of cinnabar ore, for the appearance and color are almost entirely the same. With this, it is shown not only in the daylight what the material is like, but also it indicates what kind of stock we should search for and what kind of metallic seeds we should find. Note well, the gold and silver produce such seeds, which subsequently tend to grow and to increase. In Arrige Aures Pamphile all these words are noted and confirmed by Berhardus, Geber, Alani, Raymond Lully, and many others. Now, this is enough of the review of those works, the material from which we shall take.

If we now wish to proceed further, we must now go on to Paracelsus' School of Work and to the brief passages which follow, as are readily to be noted and learned from his Scriptis. Then we desired to learn from the Subjectum Materiae vel Tincturae, which is reported clearly enough and shown with the fingers, so he also set up thereby the welcome preparations and said that from one (of the materials noted) two would result, namely, sulfur and mercury, since one of them (mercury) gives the body, on earth, wherein we plant the seeds, i.e., the Sulfur of the Sun, which sulfur is called the Blood of the Red Lion, while the mercury is called the Gluten of the White Eagle. He further stated that when the two coagulated together, then the tincture thus prepared was ready to be used. This is, indeed, a shorter way of preparation, though it gives poorer and inferior results. In addition, he said that, however, this required the services of a well-experienced artisan to serve as Fire master. What a figment of the imagination!

Now, however, everyone might like to know how to make from one (of the materials noted) two different materials and bring seed from the earth and afterwards sow the seeds now and again, so that there would be an increase in the fruit brought forth. Indeed, he did not set forth any separate process of operation therefore, but nevertheless said that those who could not understand from his description would be allowed no further conjecture and, moreover, that nothing more would be added to the Art by God Himself. Thus, it is, however, true that without any further instructions, such words would be of such a limited understanding that they would be almost too difficult, if not impossible, to comprehend and would show further how one should proceed with the material. Then, alone would he be able to say that the material or the tincture should lead to an understanding of the Red Lion, that is, that it is cinnabar, and that it is a thing which involves "one in two through the art of Vulcan", an understanding of the use of fire and heat and how generated and how released from these materials, and how sublimation may be brought about, and that I pass it on to you with its correct nature and the uses of old, so that it is the Red Lion of which much is spoken and little is known. Now, the same may, through the help of Nature and the skill of the artisan, bring about a transmutation in the White Eagle and then, too, may from one (understanding the nature of the material) there come two, that is, sulfur and mercury. More than that, the brilliant reflection from the gold should not illuminate the spagyro. As he would say no more than that, as the two principles, sulfur and mercury, are not used to conduct the work, but are to be kept in reserve, as in the case of the material, which must be only one, and yet, at the same time, may be two. In this brief work is explained not only completely the material, but also the entire work, beginning, middle, and end; and the whole process of preparing the tincture and also the complete work, in which the three things are pointed out, first, the volcanic action, which was treated by the poets in the case of fire, by others as the help of Nature, and by a third group, as the skill of the artisan, and through these three means were the above-mentioned two things, the sulfur and the mercury, in the light made by the Red Lion, caused to happen and terminated therewith, saying that beyond the brightness of the gold, the artisan was not illuminated. As he was wont to say, not to bring the work past that point, either through the help of fire or through the skill of the artisan. Then, when the artisan brings it so far that he has made sulfur and mercury from the material, then he cannot go any further, so he must let God and Nature work it out and then await the fruit. Even as the farmer, who plants seeds in God's Name and in God's Seal and proceeds thereupon and God and Nature set about accomplishing the results. The artisan must also do his part if he hopes to make the two from one. What now, however, the three works and what each official or minister of Vulcan or Fire of nature does, using the skill of the artisan, I will now explain briefly here and clear it all up and further point out what is understood by these words, before, however, reporting that us as the Sun in its course through the sky, no one else than through his reagents, which are actually from God Himself, is lit up (enlightened). Men can also see that there is a reagent in this case here which ignites our Sun the same as any fire on earth, since God and Nature have provided him to enable him to work and keep himself alive. So it is true that the heavenly fire from the Sun so influences metals that it is up to us a common earthly event and involves a solidifying and congealing fire. In Heaven, however, this is resolved. If it should come to life, however, it must be through its resembling in appearance that external fire and such a uniform fire was assumed that it burned with a bright flame, which was the spirit of the Fire, and it caught fire and had its effect. Since, however, we are not able to have the Power of heaven or even bring it down to us, then we so needed the Materialistic Fire, which was originally ignited by sparks, so that it burns continuously without ever ceasing and, as long as it has this same material making it up, and can continue to work and by means of the same also acts in metals, due to the igniting of an earthly sun, just as also like can be brought to life by like.

Thus, even if wood contains the Fire Principle and is of a combustible nature, still it does not catch fire unless ignited, as by sparks. Moreover, since this fire is not that of the shining Sun, nor is it that of shining gold, it must then be the result of the earthly fire, which is ignited by its sparks, so that it also burns and continues to burn so long as the same combustible material is added to it, so that the one Sun and Light ignites the other, just as the Sun in the Firmament of heaven ignites all the other Stars. This declares to us, as Theophrastus first noted, by way of volcanoes or fire, which the poets have termed the God of Fire or the Smithy of God, through which is understood the materialistic fire in our oven. However, natural Fire, and Nature herself, has in part been suitable to this work, that is, the seeding of fire, which so influences the metals, through the heavenly impressions and markings that were brought about through our fire. And since we here in the world have doubled the heat in our work, so we also are able to complete our work in shorter time, so that in many years Nature is not able to do it at all. The work of artisans is not otherwise than tending fires and managing the same, so that the fire will not be too large or too small, and this is the Key to the whole work effort. Moreover, the sayings of philosophers are filled with this sort of thing in which the whole of the Art consists. Fire and Azoth are sufficient in themselves, and this is enough to say to you and, as various sayings evening the works of the philosopher Bucher have noted, putrefaction, rectification, calcining, distillation, coagulation, sublimation, incineration, etc., lead thereby the foolhardy to conceive that it is everything under the heading of the above-mentioned work, which, nevertheless, since it is our own, should be considered, in our own words, as dictum: "Solvate, coagulate, and pour through a filter bed of rocks, which results in the conferring of considerable success". And, by this work, you will completely fulfill the sayings of the sages, which runs: "Bring Salt to Metal without Corrosive Action and Danger, Thus will you have White and Red".

And now, as Paracelsus has set forth, the entire operation of the separation of materials, and as you are about to make two from one, by adding only fire, so teach again the same two components and say that one should coagulate the blood of the Red Lion with the Gluten from the White Eagle, whereupon the entire tincture will be ready, thereby showing that the entire Art must lie in a single work which will be initiated and completed by a single thing, from which it also has the designation, "By Universal rules, correct and strict". And as the work must be completed in an oven and a vat, and with instruments, nevertheless with the changes in the fire counteracting the secondary effects. And also, the materials, once they are included subsequently in the convenient instrument appropriate thereto, the name of which is not given, until the tincture is ready, therewith once again if is full of the language of philosophers, as follows: "In a vat containing the instrument, you can carry it out from beginning to end". Whoever has diligently studied Bernhard's Parable, which states that as the King goes alone into his bath and takes none unfamiliar with him, and as long as he remains therein, until he comes out a blood-red color, and also like all the others, the meanest man on earth may be his protector, while no other trouble demands further attention than to heat up the King's bath, he who looks after all this diligently, will understand my meaning very well and also understand how my words are in agreement with those of philosophers.

Here I call to your attention that you bear in mind what I have made known to you in my previous universal work and how everything in itself has its own seeds, therewith to propagate itself and reproduce its own kind, and how the Red Lion, which is cinnabar transmutes itself, according to the words of Paracelsus, into the White Eagle, "only by adding fire to putrefaction and sublimation into wholesome life", and how, at first, the black color appears, and thereafter the white, which is indeed the proper White Eagle, actually mercury, and at the bottom of the glass vessel remains the blood of the Red Lion and it is the "Sulfur of the Sea", which will be extracted through its own mercury, although in the same place distinct additional mercury will be realized, to work with, to solvate, and to coagulate, which involves the older and longer process, so that he does not know thereby the proper directions, can overload the seed, flood it, and suffocate to death, so that it can bring forth no fruit and is no longer, therefore, able to carry out its function. Accordingly, so must the work be begun with wisdom that the red tincture will be obtained, then the white material must be divided into two parts, one to augment and the other to reduce, the red. In the short method of production, however, the work goes faster and more safely, since all the material involved is kept together and treated at the same time, and nothing therefrom is obtained thereby unless God and Nature have ordained it. Also, it takes mercury, the matter in metals, which is mineral, the semen of man, which is sulfur, themselves added together, with nothing more, that is all that is necessary, so let the powerless alone, for in this way the worker in the field cannot easily be deceived, since he has only to look after the fire. We have here a good example of the evolution of man quite gradually, as the man and the woman often attempt to have a child without success, but only at a time when in the first conception the man's seed is productive, while at other times all of the other seed is lost. Success occurs only through the natural heat of the woman's body, which matures the seed of the man and finally, through the interaction of Nature a fully developed living child is obtained. The warmer the child and the better the mother of the child provides nourishment for the child and carries out her other daily chores, the more perfect, the stronger, and the healthier the child will be and the mother will remain.

Here I also have, from the natural understanding of God's Holy Word, and from the evidence gathered by ancient philosophers and written about the actual process, so that it can be quite clearly recognized, noted that it can be quite clearly recognized, noted that it is not otherwise than that the proper subject, actually cinnabar, be taken, which is beautifully cut like a ruby, as well as in Lecture vz. 8 c., which is found in the mines of the Hungarian mountains, in which place Paracelsus has also shown us, saying that for our king of work no better material can be found than that from Hungary and from Istria. Moreover, on Spruce Mountain, near Gold-Granach, where no other ore was taken if it did indeed yield up gold, then it would be adulterated and impure and not suitable for this kind of endeavor, insomuch as the same generally represented two metals in one or is, on the other hand, contaminated with other metallic "ghosts", so that when this work is completed, a part of this well-prepared Tincture of Natural Philosophy is taken, along with many hundred parts of each of the imperfect metals, which gave the same a more pronounced tincture of real gold in an instant than did the natural and so eliminated the entire examination, the report, and even all other illnesses that medical doctors were unable to cure. However, they were restored and then introduced by a new youthfulness which a 60 or 70 year old man will also see, which will make him feel like a 20 year old youth and many more wonderful blessings than this, which have the ability to act as a strength-giving medicine. And how such a tincture is to be produced in greater amounts and how it is to be administered, and how projections and proposals can be brought forth have been reported by me earlier.

One also sees how the work is brought to completion from a material in an oven, glass container, or other type of receptacle, in one experiment after another, by using increased amounts to heat. And, even though the work and the art by you yourself is completely inferior in quality, the philosophers have also pointed out "It is wife's work and children's play". Even so, it still requires a well-qualified artisan to be the fire-master. Moreover, the work may be completed easily enough without danger in 16 weeks. The entire operation can be described briefly in the following verses, which were initially penned in Paracelsus' own handwriting:

> There is a stone, and yet no stone
> In which the entire Art stands alone.
> It is found under the ground and mountain,
> Wherein dwell giants and dwarfs.
> I tell you this out of kindness,
> Whoever has the same thing, the mist,
> And the Red Lion shines clearly,
> Drawing out Mercury, completely, pure and clean,
> Likewise, in the same manner sulfur, and I tell you truly,
> Which serves as the basis of the Art so completely.

I must report still a further word to you, for above it is said that no other one should be taken than this one, which one wishes to attempt to transmute into gold. So know also that all metals arise from a common root source, as pointed out sufficiently, but still one metal or ore may be found purer in Nature than are the others, and on account of this, since one of the materials mentioned could not contain cinnabar. It might even be a copper itself, extracted from a pure ore which comes the closest to transmuting the gold, as earlier pointed out. That such ores were themselves alone mentioned, which were worked and then combined, must be added subsequently as a pure seed either to gold or silver, for by itself alone, however, can the mercury, as the philosophical soil, produce no seed nor sustain any growth. Then the philosophers say: "Our stone does not tincture anything and must then have been previously tinctured itself, only poorly completed, clear, snow-white, as in an ore tap, like a wax on a soil, which when well fertilized and pressed against, goes forward. Then, when you come to think that all metals come from a common root, you will soon be able to exhaust the subject of the origin of minerals".

Nevertheless, I shall not conceal from you the fact that the mercury cannot be extracted from one mineral ore as well as from another. It is very easy from the cinnabar. Moreover, if sulfur is burned, from the silver it can also be readily extracted, and also by gentle digestion using moderate heating and from other ores copper, antimony, iron with stronger heat, though still very impure, iron alone extracts lead, tin, gold bearing pyrites, gold-quartz, all these with stronger heat, and solely from mercury ores or minerals, but most slowly of all, then minerals containing the noble metals, sulfur, resembling a ruby, apparently bearing it itself.

Now, will we, in the name of the Holy Trinity, with this passage, and sufficient philosophical advice, end this and write the actual laboratory report.

Resume of the Theory and Practice of the Whole Tincturing Process

Take the raw material, which you readily recognize as entirely pure and, as such, has become most beautiful, since it is not contaminated by any other metallic, dark-making "ghosts". Then, when dark and not pure, so that as earlier pointed out, it might not be safe, so that such purifying must be carried out in order that the work will not be ruined. In case, however, a question arises, although no particular test gives positive results, which could be identified as such, still it is much safer to work in Corpore Lunae (the Body of the Moon), that is, in finely powdered silver, since after purification and fine powdering, it is distilled "by

84

Saturn", so that no strange spirits will be in want
of anything, and in its original manner, as mercury,
which is in the first material reduced by mercury of
the philosophers, which is purified through one of
the mercuries, as pointed out earlier, and from a
beautiful, clear, transparent ore, extracted and
made pure, and also of a crystalline nature, which
is a special secret of all secrets, with which one
could work more safely and not be mistaken, the
grasp of which they know very little, but Paracelsus
had discovered such and he then wrote with clarity:
"He knows the destruction of the metals as well as
their reproduction". To bring this about, then, the
older philosophers had many kinds of ways of
grasping it. They also had many kinds of materials,
one of which is even suitable in ordinary water and
is the slowest of processes. Then there is the
process of Basil Valentine in copper, of Bernard in
gold ore, of Roger Bacon in antimony, and those of
many other philosophers working with other minerals,
although they all have an effect on one another for
a purpose (although they all react differently),
they are in exact agreement, since you know now how
to treat it. I have discovered here a remarkable
secret (in other manuscripts where the mercury
extracted from the mineral shall give to it --- the
extracted material --- its form and shape through
its own characteristic properties, which I will
never disclose. Moreover also, since I have your
attention, I will say that the extraction from the
mineral cinnabar reveals (and I have also delayed in
telling you this) that the red material or carbuncle
should be dashed to pieces and then finely
pulverized, like its own Corpus Mercurii. Then when
that does not occur, the mercury cannot afterwards
dissolve its own corpus (substance) and this is

likewise the greatest secret of anyone in the entire alchemical art and when that does not happen and the mercury is barely extracted by itself alone in a colorless form, it will nevertheless go into solution. For that reason I have disclosed to you such information in the strictest confidence, because the sorrow of Christ ordains it, for it is the greatest secret of the entire art of the alchemist, so one must have the acrid water which is able to dissolve metals, whose characteristics you now know intimately and are here distinguishable on the basis of the administration of fire and leaves its mark on all fleeces, with which indeed no mistake could possibly be made, the certain way to bring about transmutation. Nevertheless, both methods should be known, the old slow way and the new fast way, but the latter is the better way.

For this reason, consideration should be given to material cinnabar which is absolutely pure. It should be pulverized as finely as possible, until it is as fine as meal, using a marble mortar and pestle. Take 12 parts of this same powder and mix it thoroughly with one part of slightly colored Mountain Luna, which is increased silver by nature, which is dissolved by the aquafortis, then precipitated, and then is completely sweetened by rain water, which scours the crude red material very well all though one another, so that one or the other might not be recognized very easily, and then they were placed together might not be recognized very easily, and then they were placed together in a vial and luted until completely stable with Luto Sapientiae and allowed to stand in a sand bath for one month, so that only the mouth of the flask extends above it. After 8 days over a gentle fire,

so gentle that one can endure quite easily the
putting of his hand in the sand without being burned
and then for an additional 8 days and nights over a
stronger fire, with the temperature rising only
about one degree, and then finally during the last 8
days with the sand so hot that a drop of water
evaporates immediately therefrom, while the product
is making a hissing sound. It must not be allowed to
become red hot, however, for when this first starts
to happen, quickly take it out, and you will then
find a white material at the top of the container,
which in reality is the prime material, or the
mercury, which is the living soul of the metal,
through which hereafter all corpora of the metal can
be reduced to the prime material or Humidum Radicale
(i.e., radical containing moisture), which is the
"living mercury" of the philosophers. This is the
Secret Water, which has indeed given so countless
many names, since all of the philosophers have
written so much about it, yet it is still so obscure
that one must admit that it does indeed, hold in
very great secrecy that which God wills.

Also to the Glory of God the Almighty, you have now
the preparation of the clear, strong "vivifying
water", which is indeed the "Acetum of the
Philosophers", in which all metals are converted
into their prime material, the Humidum Radical or
the living mercury, the starting point for
alchemical art and work, the masterly essence of
living matter, I will now explain to you how you
should purge the mercury, or separate the pure from
the impure, so that no filthiness will be present
that might bring shame to you, whereby our mercury,
so sensitive and pure, might be added to its body of
matter, and the bridegroom lie with his bride on

their wedding night and make her pregnant, so that their family might increase and children be born without number, but, then, when this purification of the mercury does not take place, it might be impossible to finish this work. How this is accomplished, I will subsequently explain to you in complete honesty.

When now you have also extracted the mercury and it occurs by sublimation as a beautiful white product, beautiful bright, and shining like a crystal, then take the same, completely purified, from the vat and be careful with it so that no impurities are added. Then take the mercury and pulverize it in a clean mortar and pestle made of marble, causing the mercury to move gently. Note well: Repeat the process in a hermetically sealed vial, then again set it in a sand bath and allow it to stand over an appropriate fire for 2 hours, where one could be able to allow to remain for the entire 2 hours, and therefore heat for an additional 2 hours in a more drastic environment, and still further for a third 2 hour interval at even higher temperatures over so strong a fire that the sand bath makes hissing sounds when water is sprayed over it. In this way, the mercury was prepared from the silver and its excess in the silver overlapping and remained as such, whereby it is appropriate to extract the soul from the body, for this mercury is the spirit which is so fiery and draws t itself the soul from the body. For it is indeed the spirit which is the vehicle of life. Now you must see how the elements were separated. First, the Spirit on the mercury of the philosophers was drawn out from the body in the first preparation of the silver, after which followed the soul, and lastly, remained the dead body, which became ashes. Also you will need to see how the Spirit was united with the Soul and you will certainly see a great wonder in the above separation. For rays will shoot now and then into the glass vessel, and the peaks of the rays are swept together like long spears, and above each of the spears, is a natural star, which the hands of man do not have the power to make, and this is the new creation of the new world, or, as the philosophers would say, is like the birth of a new world, with stars like the colors of the rainbow. This secret belongs more to God than to man, as Paracelsus said. Then we see the Purified Soul, which will also illuminate our own soul on the Say of Judgment, so that it will be purified from sin and be perfected, and then you will see a completely beautiful thing, the wonder of which may be like nothing ever before known.

At this point we will write further and look about to see how the body will be dissolved and the Spirit of Life drawn to itself and received with joy. But, nevertheless, I will here and now explain how this mercury shall be enlarged and then will be able to permit many of them to the washing of the black earth, as follows: Note well, that here we think how this mercury was made, so that at this point only one example of leaven, as it is added to the flour, so will leaven be added to that which we use in our alchemical art.

Increase in Mercury

Take 12 parts of this purified mercury and 1 part of finely filed silver and from them make an amalgam, which will also be finely pulverized and put it into a vial and hermetically seal and then allow it to stand for a prolonged period of time in warm water until is noticed that the entire body is converted into mercury, which becomes corroded away, like aquafort does to the silver, and also as it does to all of the mercury, and in this way you can increase it forever and even ad infinitum. This is spoken of as the "whitening of the black earth".

Solution, Putrefaction, and Mortification

In the Name of God, grasp and retain our digested preparation of mercury and take one part of it in a flask or vial and mix it well, then set the container in warm ashes for 6 weeks, whereupon you will observe precipitation taking place at such heat, until finally there is an ending of the intimate contact, and at the bottom of the flask a black substance will remain. It will not be black as coal, but will be dark as ashes. Then you will know that the first phase of the work, the solution and putrefaction, is complete and that the mortification, is complete and that the mortification of the body is actually taking place.

How the Black Earth is Made White and the Dead Body is Raised Up From the Dead

Treat the black earth with a half part of mercury, prepared as above, then pulverize the two materials together in an appropriate glass vessel, making certain that the two materials are thoroughly mixed until they are subsequently in intimate contact, which process requires approximately 8 days. Thereupon, you should observe the Soil, whether it shall be transformed from the black color to the white or remain blended together. Then take it out and treat it with mercury or the silver water (nitric acid). Seal hermetically after pulverizing and mixing thoroughly and again allow it to stand for an additional 8 days. Then take it out for a third time and allow it to stand again with 100 grams of nitric acid with diligent stirring to achieve thorough mixing to allow it to dissolve by intimate contact for 8 days and night, and then coagulate as before and treated with 140 grams of "silver water" (nitric acid) well triturated together and mixed and then allowed to stand for 8 days and nights and then allowing to coagulate, At this time the other work is also complete.

The Partitioning of the Stones as the Third Phase of the Work

According to the phase of the work you will be partitioning the gem stones and this involves the use of a white and red tinctures and shows you at that time the earth, which before was black, has now been whitened through the action of nitric acid, so that it is prepared to receive the entire ferment or Soul, which is white or red, and everything comes to pass according to the rule that what one sows, one also reaps, if one sows gold, then one reaps gold; if one sows silver, then one reaps silver, and so, when one sows gold or silver in the freshly plowed earth, then he will bring forth in a hundredfold yield.

What the Ferment Should Be

The ferment or seed, which is allowed to lie in the soil should be a very beautiful, carefully prepared silver-lime for whitening, which should be porous and white, as I will speak more about later on, and for coloring red it should be a very beautiful, carefully prepared gold-lime and should also be added, together with the previously mentioned amount of material, checked subsequently, so that it consists of 21 drams of white earth, 14 drams of moon water, and 10 drams of lime or ferment. It is also to be understood that for 3 parts of white earth there needs to be 2 parts of nitric acid and one and one-half parts of Ferments. These three items shall be finely pulverized together with special diligence in a marble mortar and pestle, along with good mixing taking place at the same time. The process is carried out in a glass vessel, well-covered when the material is added hot. The material thereupon coagulates into a hard white material, which is removed and pulverized finely and then added to the third part of its water and subjected further to moderate heat, whereupon it begins to sweat until this liquor is evaporated and the product becomes a stone. This evaporation, and the diligent effort it requires, should cause no annoyance, since our stone is enlarged as a result and in this way becomes improved and perfected. Thus, the imbibition is allowed to take place many times, until the stone becomes complete and fixed and all defective material is tinctured in the Lunar Liquid, or gold, to the Glory of God.

Careful Markings of Merit

One must examine and pay careful attention to
whether it will melt on a hot copper-lead block
without fumes, but where this is not the case, it
must be treated more drastically to be dissolved on
standing and therefore coagulated on a warm ash bath
and dry on a stone, whereupon it is ready to be
tinctured.

The Stone to Use for This

When this work, as earlier pointed out, is finished,
can take the first white stone and divide it into 3
parts, the first of which is kept for preparing
future products, which shall originally be brought
into contact with the Ferment and then well-dried on
the fire. The drying process involved here can be
carried out in two different ways, the first of
which allows the material to stand for 8 days and
nights and thereafter be further treated with
activated mercury (first of all, however, purified
with the silver cleaned mercury or silver water
(nitric acid), 3 parts, triturated and imbibed, as
earlier noted, where it was stated that the same
would be needed to convert the black earth into
white earth, and when imbibition has taken place,
then sweating (syneresis) will occur, so let it, and
take the liquid by itself that the entire process
may be maintained and expanded, like an eternal
work, in this way from an initial 3 days and nights
to 8 days and nights. On the other hand, when you
have the white earth ferment, which is a sizeable
undertaking in itself, you are able daily to
accomplish this by using activated mercury, which is
added to "silver water", 15 parts, and then allowed

to increase by itself on standing from day to day, the weight calculated to be also 15 parts. How many parts one takes therefrom depends on how much one wishes to risk. One might wish to risk the excess, which one knows to be the 15th part of his silver water, which he has added from day to day, first by adding a few drops and rubbing it in, then as the silver water becomes well mixed in, heat carefully at an absolute minimum heat.

Another part of the whitened stone will be used in the fermentation process, involving silver-lime and its water (nitric acid), which activated the mercury and works continuously, so that it produces a perfect tincture, as is described abundantly, and satisfactorily, in a previous work.
One may use the third part to prepare the red tincture, which then lies hidden secretly in the white tincture, which is recognized in the work.

How the Red Tincture Should be Employed

Take one part of the white tincture, which had been heated earlier then completely pulverize it, and put it in an appropriate container and heat at a moderate rate (but greater than before in the case of the white tincture) and it will be sublimed within 14 days. The sign that you will especially take not in this part of the work is that the white earth is turned into red earth, which then increases, depositing itself on the glass walls of the container, so long as the fine is continued and the heat maintained, it all becomes red, like burnt saffron, and the earth is ready to receive the seeds of gold, with which it first shows an increase and then shows imbibition. Here, it happens, as the philosophers have pointed out, that the red man marries the white woman, so now the red earth, which previously was white, is taken and pulverized until completely fine, and then wetted by the Mercurium Solis (royal mercury, i.e., activated mercury); which is prepared in many forms, like the mercury of the moon. The Mercurium Solis (literally, the Mercury of the Sun) is prepared, as follows: Take mercury which has been thoroughly purified as previously indicated, and 10 grams of very thin gold leaf, beaten in a gold-beater, and finely pulverized, and from this prepare an amalgam, as previously described in the case of Mercurium Lunae (Mercury of the Moon), and put it, after thoroughly mixing, in a vial, seal it, and allow it to stand in warm water, until you see that the gold from the mercury which has become corroded also increases the mercury from the Sun. After moistening this product with the Mercurium Solis, the moistened red earth is now made into a stone by rubbing the resultant material together intimately, then pouring it into a glass flask and allowing it to warm for 8 days over a small flame, during which time it will undergo syneresis until it dries out and becomes hard. This entire operation is repeated a second, third and fourth time and each time the process is carried out just like the first, including the wetting down, the addition of the charge, and allowing to undergo syneresis for 8 days and night. After repeating the fourth time, then it will be time for the stone to give up its Soul and undergo fermentation, until a red product is obtained, which is then weighed until the earth amounts to 3 parts of the same, the gold water, or Mercurium Solis 2 parts, and the gold-lime 15 parts. The resulting product is then thoroughly pulverized and the well-pulverized material is treated as previously. The amount of heat is slowly increased and as a result of syneresis and breaks up into smaller particles, it reaches its final form, from which one part is taken as the increase and treated as before by Augmento Lunae. To redden it, Mercurium Solis (Mercury of the Sun) is used by itself alone, so it becomes excited by the gold, whereby the Trinity, body, mind, and soul, again come together and then appears to be composed, here in the red words, again indeed of black, white, and red colors, and even in the white works, from which you shall be able to observe the progress of the entire operation with the greatest of diligence, as has been described earlier in the work, from the preparation of the white tinctures up until the completion of the entire operation, and the liquid had been used up which had been fortified with gold water (i.e., choice Danzig brandy) in an equal amount by weight, as previously has been noted in the case of the white tincture or silver water (nitric acid), and as so often was done, again moistened and soaked until the final completion, as also was noted well, that you can maintain in the red work the fire at such a heat that thereby you will obtain the stone colored throughout in the deepest red color, like that of a garnet. Thus, by the Almighty and Compassionate God, everlasting Holy, Honored, and Praised.

Exhortation to the Gracious Elector

God Almighty has made himself to be a judge between Your Gracious Elector and me, so I now report the truth, for your Gracious Elector will do me good or ill on the basis of my deeds. I declare this, with God's help, that this is the only clear basis and

knowledge of Holy Works of Art and Mysteries, of
which I hold back nothing, the comprehension of
which alone is impossible for me to describe,
although Your Gracious Elector himself,
nevertheless, has adequate knowledge. For when a
thing is indeed represented so clearly, and when it
is undertaken, however, it may (as so often happens)
proceed under other circumstances. Therefore, since
I have such a significant witness, neither I nor
your Gracious Elector can be deceived. I must also
take care where I should like to do it, even though
such might be punished by God, that in revealing
this divine secret of God regarding Your Gracious
Elector of the salvation and the holiness that you
can allow to come to this place, for the purpose
that it will make good use of what is otherwise
evil, and, for this, you must, therefore, give the
strictest judgment of God's reckoning and of whether
Your Gracious Elector believes that it might be a
lie or a fable, for so shall you know,
notwithstanding, that I do not spend my hope of
heaven so frivolously that I do not report the
truth. This, then, like everything else described
here, and touched by my hands and seen with my own
eyes. To me also, in other respects impossible and
inhuman, in this practice, reveals the influence of
the color, the time, the day, and the hour, as well
as also of the entire work, the material cinnabar,
the solutions, the coagulations, the white and the
red work so ordinary in its execution and in its
reporting, nevertheless may be, in no small part,
philosophy, as long as the world lasts, and so, Your
Gracious Elector, in no philosophical book ever made
can be found the time of your life, or can it even
hope ever to be found in the future. Meanwhile,
since I am baptized in Christ, as well as any other

Christian, as the Great Lord and Almighty well
knows, I am also mortal and all sins are brought
into subjection, so that I do not know when God will
call me Home, whereupon the elements will be able to
be separated from one another, but I might not be
involved myself or be able to do this, since I do
not wish it.

Now, if Your Grace, the Elector, please, so dear to
you is your happiness and gratitude to God in Heaven
to permit you to understand that you are worthy of
this most important secret, I will reveal to you
that which is the highest and the best. I will also
not hide from you the slightest detail, but will
reveal to you and explain to you the miracles and
wondrous signs of our Mercury through the threefold
paths which are to follow, so, accordingly, I myself
appeal to Your Grace, the Elector, at all times, to
demonstrate so much in 4 weeks that it can be seen
that my suggestion is valid and not the result of a
sophisticated grasp of the issues, for not, as it
were, in nature and Art, however, can it be see what
our Mercury could be after extraction, provided it
is made as previously noted, namely, also, that it
is white, clear as crystal, and easily accomplished.

First Miracle of Our Mercury

If Your Grace, the Elector, please, when our mercury
appears, as a clear, brilliant, and transparent
crystal, then the same (not more than 10 grams) and
1 dram of gold wax, and 1 dram of benedict's soap,
placed together in a flask having no cover thereon,
and a gently fire beneath, then the produce will
begin to turn black and coagulate on the mercury and
it has become a metal, since it clears away the

smoke previously in the fire, although there is no metal that appears to be of the same type, in either form or behavior, still however, when it has become a metal through the coagulation process, then it appears as the most beautiful silver, although it is not at all clear at this point that the real cause is whether it will take up sulfur itself or not, namely, from the wax and the soap, which are both fatty and sulfurous, but this information is not exactly familiar and well known and contrary to its own nature, since it requires nothing other than its equivalent of sulfur to coagulate it with, and it has also coagulated this sulfur, as Your Grace will now see, for it is nevertheless a property of this sulfur, but is instead a beautiful wonder of Nature that transforms our own mercury, through a less understood principle of nature, in its form, its content, its properties, and its characteristics, into a metal. Be that as it may, it is not, however, like other metals, either in form or in content. From this, Your Grace, you can conclude how soon it should be performed, when something foreign comes into our material or something impure might result from our work. At this point, Your Grace, since it can be accomplished through the help of trusted friends, as previously reported, in seeking the pre-determined Hungarian material, which is entirely pure, although whether indeed to obtain the same, which would require considerable expenditure of time and money, and to introduce all of it into the work and thus advance the progress of the world, might be debatable, even though it would make the artisan less confused in the operation of the process. It could also be carried out quite well with material which can be found in Germany in some of the mountains around here, although this has not been

found to be necessary for a long time, since indeed
it frequently could be the cause of spoiling the
work, on account of the many impurities and
extensive types of minerals which they have there,
which hundreds of specialists in the field do not
recognize or know how to analyze. Consequently,
then, many of these may constitute serious problems,
on which God's creations, Mother Nature, and the
writings of the philosophers are looked as the cause
of such inaccuracies, although the more reasonable
men are able to handle such rough understandings.
However, when one is not illuminated so broadly by
God, and the work does not reward him according to
his expectations, then he should, nevertheless, not
disdain t desire the lofty endowments of God
concerning his ignorance, for which reason, Your
Grace, you must take this opportunity to be reminded
constantly, to be sure, and permit yourself to bear
in mind the great wonder which the Almighty God has
stored up in these materials, although this is a
situation, at least, you will not long have to
endure.

Second Miracle of Our Transmutation of Mercury (Copper into Silver)

On the other hand, and even still more and most
wonderful of all, when I take 10 grams of this
copper and add it to the molten metal, then the
copper is colored with such a stable color, which is
like that of the silver, so that also this white
color of the copper may not be taken up by any
route, since even the copper is driven off by the
lead and it is then that the copper is driven away
completely by the addition of excess lead
(nevertheless 100 times as much lead must be used,

and before it is gone, the calculated amount of copper is added) thereto and the wonder is observed and our mercury is stable and in the fine has its strong coloring and penetrating power, which can always be seen here and should be noted more fully where its characteristics are observed for a longer period of time and Nature's laws are followed more carefully.

And, Your Grace, when the metal mercury was treated likewise, like a large bean carried in the flow of molten metal and so was converted instantaneously into metal, which disappeared and could be seen no longer when it came out, without any smoke, just like fire consumes tinder or wax melts under a light and becomes soft as a result, so does ordinary mercury assuredly coagulate in the fire and smokes, although gold and silver are therein, which then is a wonder of wonders, when this is also combined with a foreign metal, which is unlike it and can be sufficiently recognized, since it is a wonderful display, which is not like anything else in Nature.

Third Miracle of Our Transmutation of Mercury (Mercury into Silver)

In the third place, Your Grace, from our mercury metal, I must tell you further about the wonders that take place when you take the metal mercury and pulverize it finely by filing silver and putting all of the filings in a small flask and pouring over them carefully distilled spirits of wine, so that the filings are covered to a height of one finger. The liquid portion is then poured off and the process is then repeated several times, the more the better. The wine drawn off is finally heated, whereupon the mercury melts. On cooling, the residue of molten metal contains the mercury, which at this point is an especially beautiful white color, like silver, and this material is then pulverized, so you will be able to find out how much good silver has been obtained from the mercury, sothat it at once may be evident what was the result of our sophisticated study, not to mention also its proper carrying out, so that Nature can thereupon take its course of following up.

And now, Your Grace, we entrust further developments to Almighty God.

Other Universal Work and Short Summary of the Lapidis Benedicti

Take red-golden ore (which may be contaminated with small amounts of foreign material), which is beautifully transparent, much like rubies, which are found in the St Laurenz and St Vincenz Mountains of Hungary, which were referred to by Paracelsus in his Tinctura. Because of this, no other ore was even considered, since it behaves like gold (and contains only Lunae Spiritus) and is volatile, but actually this ore is as finely pulverized as dust on a marble block and not on metal, and remains pure and spotless. Weigh out 12 parts of this powder, mix well with one part of finely filed silver, and then add the silver-chalk from the Aqua Forte (nitric acid) until precipitation occurs and then wash the precipitate well with rain water. This does not permit analysis in this work, since it is only opinion, that is to say, arrived at only by thinking about it. After grinding well together, the product is carefully sealed in clear-glass round-bottom flask, and allowed to stand for 24 days in an ordinary circular bath or 42 days in a dug bath, so that in 14 days the Head of the Raven appears, and the same should be increased gradually, as follows: "Fecundum Gradus" (fruitful stage) of the universals described heretofore, even to the white mercury above, sublimed white and beautiful, seen as a crystal, this same mercury being then the mother of metals and the rich earth, wherein our seed must be sown. In his work, the seeds of gold remain at the bottom of the flask. Thereafter, the flask containing this material is invented so that the seeds of the gold or the sulfur fall to its own ground, in which it previously was growing. When the seeds are then swallowed up by the earth and the mercury itself is again growing up, the flask must again be inverted and continue to remain according to this arrangement until the earth surrounding the seed begins to change color and a beautiful green grass puts in its appearance and undergoes further change to all sorts of different colors, until finally the entire field takes on a golden color. At that point the work is complete. Then as often as the mercury continues to grow up because of the seed or the sulfur and suspends itself from above, just soften does it take on another form and color, which becomes very obvious. Then where the red color shows up ahead of time, the fire has had its greatest effect and must be recognized at all times from the color, for the fire must always be under control, and this no one can ever predict, since this puts him in the position of having to have his eye on everything. In such a work, all colors appear which can be found on earth, thereupon the flask frequently give the impression that it is covered over with a golden coating. However, the color disappears with time and the stable white color or Alba Regina (literally, White Queen) of which one part per countless thousand parts, remains (in all imperfect but nevertheless purged metals of highest purity), in which most of the purest and best metals are tinctured. The better it is, the more natural. Moreover, such occurs instantaneously. If one continues further, however, until the white form is converted into the red, and finally into the ruby-like type, the red carbuncle being converted into the highest type, countless thousands of parts of every imperfect metal are purged instantaneously into the highest and best gold, and this occurs in all samples.

David Beuther's Vitriol Process

When the beautiful Spiritum ex Vitriolo (vitriol spirit), next made in the flask in pure form from the gold or by heating 10 pounds of vitriol and thereupon poured off from this spirit, so that it would be well to moisten it, seal it, and then allow it to digest 40 days and nights over a gentle fire and distill over a small flame, the resulting spirit is called the Mercury of the Philosophers and that obtained after sublimation and distillation from the ore is the Sulfur of the Philosophers.

Take now the ore remaining as a residue in the flask, powder it finely, and treat it with the liquid drawn off, and then distill the product over a gentle flame for a period of 5 hours, until a white vapor is seen to arise and hang above the flask, then keep it well sealed. Repeat this process by decanting off the water extract and sublime the white material until no sublimate is any longer

obtained from the Caput Mortuum. The sublimated of
white sulfur is retained and the waste ore, or Caput
Mortuum, is discarded.

Thereafter, these white fumes of sulfur are sublimed
one more time and the waste material remaining
behind is again discarded. Take the sublimed sulfur,
as is, from the water distilled off and the mercury,
as is, and put them in a flask, seal it well, and
heat over a gentle flame, whereupon you will see an
increase in the white clouds, which will be allowed
to stand until they have coagulated. At the bottom
of the flask, everything remains dry.

Now take just as much of the above-mentioned water
as is equal to the weight of coagulated sulfur, mix
thoroughly, lute well, and allow to dry as before.

The third time, half as much of the water is added
as is equal to the weight of the material present in
the glass at this time and allow to evaporate until
coagulation takes place and the product becomes dry.

Fourthly, add half as much of the water as is equal
to the weight of the material remaining in the flask
and allow it to evaporate till dry. After such
procedure, treatment of the material is considered
to be completed.

Taking the coagulated material, lute it in
"philosophers' egg" until solid and then digest it
gently, until the most beautiful colors that you
have ever seen appear one after another. Ask no
questions about anything that follows, but allow the
mixture to stand until it appears as a black mass,
then do nothing further until the black mass becomes

as white as pearls, then heat slightly more vigorously until it begins to turn red, then keep heating a little more until it becomes blood red, then turn up the heat as high as possible, until it can no longer be turned up.

Take up the powder, as prepared, in wax and add it to 100 parts of molten gold, whereupon it will be pure tincture, of which 1 part per 100 parts will purge copper in the flux to tincture clear gold.

Ammoniacal Process Carried Out by Beuther and Colleagues at the Royal Laboratory

On November 1, Beuther and I each weighed out 2 pounds of antimony, 1 pound of iron-filings, and 1/2 pound of prepared chalk into separate crucibles and put them in a drying oven. The crucible was covered with flat charcoal and the oven contained well-covered charcoal and was allowed to stay at full heat. Thus the antimony was added to the iron, and the chalk to the several parts which were, however, impure. When all this was covered over, the antimony began to flow into the fat. Using an ash-colored flame (having a slightly greenish cast) below the crucible and arsenic in the fumes arising therefrom above the charcoal or the covered crucible, the fumes became completely white, and as soon as it was completely consumed, boiling and sputtering commenced, so that it was evident that the reaction was taking place. The crucible was then removed, since the material therein was nearly molten and had been converted into iron, which had been concentrated by stirring the contents of the crucible, which was then removed from the drying oven, and poured into the mold. When cooling was complete, the slag could easily be removed from the finished product, which was then weighed and found to come to 440 grams, much of the original material having been consumed by the fire. Since, however, I had carried out such work in the presence, and under the instruction of D. Beuther, although I have carried out one test by myself alone in his absence and have made another reduction as a model, and have obtained the same composition and weight, and have verified the same, and have allowed it to remain in the fire half an hour longer, before pouring it off, and have obtained 440 grams of the product. Although we were able to proceed further with the reduced product, Beuther had allowed the same to stand overnight, only that he might be able to assure me how the sulfur must be prepared and used.

Preparation of Sulfur

From 2 pounds of saltpeter, in a new glass vessel, covered over with cold water, and set aside to heat, without however, being allowed to boil, was obtained a solution, on stirring; subsequently, 5 pounds of commercial powdered sulfur were poured into a glass vessel and the saltpeter solution was poured off and the sulfur stirred until it settled on the bottom and was then allowed to stand overnight. Such sulfur should thereafter be allowed to stand a good while in water and then allowed to dry. It can be obtained pure rather easily, and stored several years without spoilage and used.

Beuther's Further Communication on Antimony

When large amounts of arsenic are made from antimony, using 1 pound of antimony, it is possible by this process to introduce only 2 times as much, using iron and chalk, so that adverse conditions will be held to a minimum and so that silver can be added immediately and no longer be taken away. Also, the silver will compact and be fixed immediately after the first addition of gold, as shown.

Item: When the antimony in the fire will be too turbulent or too sparkling like a star or will appear to be too long in germinating, then pure pieces or dry iron filings must always be used. However, if too much iron is taken, the saltpeter and the prepared sulfur is added and in each case the firing or casting should surpass one mark by about half a drachma. Such a figure can often, through smelting and casting, amount to 100 to 200 grams of gold to bring the required amount of reduction. When half as much antimony is obtained and nothing is lost in the firing process, then this work has proven satisfactory. And, although the antimony, when added to the product, reacts, it is, nevertheless, very unserviceable and must be further worked to be useful.

In this chapter is now to be found the entire basis, and here we will proceed with the iron and equal treatment of the copper obtained from the iron, so that you will see what God has done for us by combining with the fixed silver, in these proportions: 4 parts of copper from iron, 4 parts of pure gold, and also of fixed silver, and then driving off the resulting lead.

On November 2, Beuther drew off the water, lying supernatant above the sulfur, and containing the saltpeter, which after standing a day and a night, the clean liquid was decanted off, and the moisture thereby skimmed off also came therefrom. The sulfur was put in a vessel and set over a low charcoal fire and then was allowed to coagulate very slowly, so that it no longer contained any moisture. Then, when it had been converted essentially into antimony, the danger might lie in the fact that it is a matter of concern.

The sulfur must be well stirred with a wooden spatula, so that it will not ignite nor catch fire. Moreover, it can combine with both substances when it is almost completely dry, such sulfur must be kept in an earthenware container, dried out completely over a fire. The sulfur can be stored after careful treatment and remain stable for an entire year and, when adequate care is taken, when sulfur is added to copper then one could not even suspect that the sulfur contained latent saltpeter.

On the same day at about 9 o'clock in the morning Beuther had his plan, according to which 440 grams of the above material were weighed out, with a fourth as much, namely 110 grams, of precipitated chalk and the same weight of iron filings, and the crucible was put over a fire until nine-thirty. The antimony did not melt as soon as it has previously, but required longer time the more difficult it was to liquefy, for the reason that it had lost its sluggishness for the most past from the first contact and when added to silver, as is its custom, it took away from the first contact and when added to silver, as is its custom, it took nothing away nothing and even immediately with gold on the addition of Luna Fixa. In another heating the antimony was somewhat more flow resistant than was the case in the first heating, even though the chalk was useful in removing its impurities. But when it had been enough in the first heating, additional amounts of chalk could be added. As often as the antimony is smoke-dried, or heated again, the usual gray-green flame in the fire can always be seen and, since it is more volatile form of sulfur, it remains on the charcoal or the instruments and stays there, completely white, and not a little unlike saltpeter. It must be stirred 2 or 3 times until it flows very readily and just as Beuther had to smoke-treat his antimony for a third time, so he also found slag, which he reported yesterday for the first time, and then added it in order to be able to bring out a new pattern. Approximately 15 minutes past 11, he allowed the entire model to cool, which weighed 500 grams after removal of the slag. At a quarter till 12, my model was completed and as the clock was striking 12, I added yesterday's slag again and put it on the fire for a third time. Both the model material and the slag now had a lead color and show small points and, as Beuther had reported, when the antimony dissolved the fixed antimony took on a beautiful gold color. As the clock struck one, I took my model material from the fire and weighed it. The material without slag weighed two and a half weights. And, while a lot of slag had been thrown on top of the model material, the slag continued to be added by the fire, so that the correct weight of the model material could be determined. Since Beuther had by this time permitted his slag a good half hour in the fire, he had a handful or three of poor quality saltpeter, along with a precipitate of arsenic which had been deposited in the crucible, The slag was stirred and eventually removed. Freed from the slag, Beuther's model sample weighed 2 pounds minus 40 grams. However, I have the slag from my own specimen, without any additives, in the crucible set briefly on the fire just prior to two o'clock in the afternoon. At about 2 o'clock I put 2 good handfuls of saltpeter in paper rolled up for that purpose and at a quarter past 3 o'clock the crucible was taken out and the sample material was separated from the slag and added to the previous sample material. The product weighed 10 grams less than 4 marks, but still 30 grams more than Beuther's preparation. This may be due to the fact that Beuther's preparation remained in the fire. From my sample material I have obtained bits of information which I can use for my model, since the sample material will be shaped according to the type of fire and so will its appearance. As the clock was striking 3 in the afternoon, Beuther had fired his sample material for the third time and immediately it melted and he found that he had 80 grams of prepared sulfur available. The antimony had been completely stirred into it. Thereby then, most conveniently, since water was flowing therefrom, he subsequently poured it into the mold. It was then a quarter past 4. The sample now weighed 40 grams less than 2 pounds. As soon as Beuther had removed his crucible from the fire, I added mine containing the sample and allowed it to remain at high heat for a quarter of an hour, whereupon 80 grams of prepared sulfur were added with vigorous stirring. The resulting product was then poured off and allowed to cool. After the fourth melting it weighed 3 marks. Therefrom I presented the Elector with 2 test sample portions in order to determine whether he accepted the silver which was offered him as well as he did show gratitude for the gold. Then on the 3rd of November, at a quarter till 10 in the morning, I fired my sample for the fifth time and when it had melted I added 40 grams of sulfur to it and poured it from the crucible. The sample weighed 2 pounds and 40 grams at this point, and 5 grams were taken from it as a test sample and added to lead in an earthenware vessel and allowed to evaporate to nothingness on a sand bath. Whoever carries out this process in a similar manner can assure himself that something must remain. After 9 o'clock in the morning Beuther had fired his sample for the fifth time and in spite of the fact that it was more difficult than my process, nevertheless only 40 grams of sulfur had been added. At about 1 o'clock in the afternoon I weighed my sample to amount to 2 marks and 40 grams. After the sixth reaction and as it was ready to pour of, 80 grams of prepared sulfur were added with stirring and the resulting product was poured off. It weighed 2 marks. Subsequently, Beuther had likewise added his sample material and, after introducing the sulfur, he poured off the product at a quarter till 3 o'clock and it weighed 3 marks and 20 marks and indicated to me that I should not be unduly concerned by this discrepancy in weight, for his sample would be heavier than mine, since his sample still had as much iron in it, although my sample might be too selective and would hold more than his, as is also the case in the test samples, until he brought his sample more into conformity. At about half past 3, I had fired my sample for the seventh time and it weighed 2 marks and 20 grams, and after serving the slag therefrom, melting, and firing for the seventh time, it weighed 40 grams, while both of them together weighed 2 marks and 60 grams. The sample at this point possessed quite coarse spikes, resembling crude stars. I thereupon had Beuther's orders and so cut

60 grams of scrap steel from screws, which was added
as a flux, 20 grams each time, quickly one after the
other, and then also added 80 grams of the prepared
sulfur, with vigorous stirring and shaking, and
finally the liquid was poured off. The product
weighed 2 marks and the entire operation consumed
44.5 hours. To this, Beuther also added his sample,
over a period of 4.5 hours, but he added no sulfur
or iron. After pouring off, the product now weighed
3 marks and 10 grams. Beuther reported this, using
in this particular process 1 mark of iron and 120
grams of flux material, and could in 10 hours of
work produce 100 quilders worth of product. At this
point it must now be noted how the process is ended
by a fixation. Friday, the 4th of November, at a
quarter past 8 in the morning, I again added to my
sample 2 marks and 40 grams of sulfur and 40 grams
of iron, added after the flux, and at about 8:30 the
product was again poured off and weighed 2 marks and
40 grams, whereas Beuther had by this time melted
his specimen for the ninth time and then added 40
grams of sulfur thereto, bringing the weight to 2
marks and 40 grams. My specimen had remained on the
10th fire for a quarter of an hour less than 11
hours and then 40 grams of steel filings and 20
grams of sulfur were added thereto. It was removed
from the fire at about 12 o'clock and weighed 2
marks minus 20 grams. Beuther had also put his
sample on the fire and it wound up also weighing 2
marks less 20 grams. My sample remained about 2
hours over an efficient flame and 40 grams of sulfur
were then introduced, bringing the total weight of
the specimen on the fire for 2 and three-quarters
hours and 60 grams of iron were produced thereby,
the total weighing 2 marks and 20 grams. At two-
thirty, I put my specimen on the fire and added 40

grams of sulfur to it. Its final weight was 1 mark and 100 grams, from which I took a small test sample. In the 13th firing my specimen was found to become molten without the addition of any further material and it weighed 1 mark 1 mark and 10 grams and afterwards both samples were melted together at the request of the Elector.

In David Beuther's process the lead was allowed to come to a boil with the specimen copper and then poured into fused silver, and the volatile material driven off.

The Preparation of the Chalk

Good chalk, which is not gritty, is broken into small pieces, the size of dice, is placed in cold water, and there it is allowed to soak for 1 hour, and allowed to stand in a new glass-stoppered flask and then set in a sealed air oven to burn until it glows and is burned out, at which time it was considered to have been properly fired and prepared as the previously mentioned model antimony sample y subsequently removing the coarse particles consisting of impurities.

The Preparation of the Sulfur

From 2 pounds of saltpeter and 3 pounds of high-quality sulfur, or 3 pounds of saltpeter and 5 pounds of sulfur, prepared as recommended at the beginning of the process, the sulfur Mars being fixed and non-combustible, and furthermore gives the proper color and the amount of fixation leaves nothing to be desired, as silver and tin. The vapors of antimony are white and red, are among the best medicines, and transmute silver into gold, that is, through the red and fixed sulfur lying dormant within it.

Some other workers may be able to purify the silver so highly that aquafort is no longer able to damage the material after the frictional wear of the lead, although only half as much product will be obtained. Proceeding with this fixed and compacted silver, as here reported, in addition to the high-grade gold, as far as is known, the silver should be free from any gold and refined prior to the test.

The spirit of antimony does not, in itself, include the substance of iron, but its quintessential characteristics antimony has the perfection necessary for a flux, while the iron gives the fixed sulfur. After these two decompositions take place, a fixed substance results. And when the antimony melts for the 20th time, with some difficulty, then in each melting the iron is a cadaver and its essence and spirit are always given in the nature and the essence of the antimony, until it coagulates into the perfect metal as a result of complete solar and lunar fermentation.

You must make use of a ferment, i.e., a substance which can remain stable, must be fixed, and cannot become volatile. Therefore, when you carry out your work on an earthenware plate or in a glass vial, or capsule, then combine it with lead but not gold, the more the better, to retain the volatile substance with the fixed substance, so that both will be fixed.

Beuther's Process from Copper Pyrites

Copper pyrites, preferably the richer ores, were pulverized as finely as possible and heated, gently at first, subsequently more vigorously, but avoiding direct contact with the flame, as when sulfur was fixed, which must not be destroyed by burning. The burning must continue until the pyrites no longer give off fumes or a sulfuric odor. When hematite powder is noted, the process is complete and the product can be pulverized in the manner in which cinnabar is. The test samples, when applicable, behave in this way: A little of the pyrites, covered with precipitated aquafort a few fingers high, is placed on a warm sand bath. The aquafort does not turn a green color, but remains clear when decanted off.

Enough yellow arsenic is now taken as fills an iron jar having 3 feet and an iron lid which completely covers it. After heating over a gentle fire for 2 hours, the heat is gradually increased until the arsenic rises and begins to give off fumes, causing the cover to rise. When this happens, the air pockets are luted again and the fire slowly removed. When the material has cooled, tapping the jar and cover causes the arsenic to precipitate. The product is partially golden in color. After subliming in a retort with attached receiving vessel, the arsenic increasingly becomes snow-white, like flour, and is also quite stable and safe for use.

Specially activated sal ammoniac in solution in clear fountain water was prepared, filtered, and coagulated. Subsequently, 1 pound of antimony was carefully fused with 120 grams of iron. When a good flux was obtained, 80 grams of calcined chalk were added in coarse pieces, which permits extraction of the ores together and separation of the slag, and the breaking up of the product into pieces the size of hazel nuts, and then pouring into a well-prepared sample of aquafort in a flask. The resulting yellow solution, poured into a third portion, and then so much of the aquafort is poured off into good strong brandy that a precipitate forms, and that is how antimony is prepared.

Finally, a crucible is ignited and 40 grams of saltpeter is placed therein. When the saltpeter begins to burn, then 1 pound of alum is added. The alum burns itself out, so that the moisture which it contained has now been driven off and this is what happens as it becomes pure and white and has been completely prepared.

Take 1 pound of the above-mentioned copper pyrites, 2.5 ounces of the sal ammoniac, prepared as indicated, 1 ounce of arsenic preparation, 1 ounce of the dissolved antimony, plus alum, and the calcined copper pyrites, steeped with half that amount of well-distilled acetic acid and half that amount of good brandy, stir them well together, so that the mixture is rather damp, then add the weighed material to the pyrites, and then when it has been well-stirred, let it stand for 4 days and 4 nights, so that the pyrites are dried out, after which time 4 ounces of mercury are added thereto. The product is then allowed to stand for an additional 4 days and nights, whereupon the mercury is lost, so that 4 ounces of mercury needs to be added again and the product is again allowed to stand for an additional 4 days and nights. The last extract was then combined with the first and as much as possible mercury is then taken and allowed to evaporate from an earthenware dish, leaving behind a residue, shown to be sulfur. To this sulfur in a crucible add 30 grams of tartar and allow it to stand in a calcining fire. The tartar was the above-mentioned white material, which left behind a grey-black material which I have termed "black sulfur", 1 pound of iron can be converted into copper in the following manner:

Bring to a boil 2 liters of water in a large copper pan and then add a handful of salt. When the salt is completely dissolved, a pound of iron filings is added and then 25 grams of "black sulfur", and in a half hour a pound (at most) of iron is added to the pure copper, 8 ounces of which were fused in a crucible and, when this mixture was completely fused, one ounce of the best Hungarian gold was added (as a seed), whereupon the mixture was poured into an ingot mold, made in the shape of a vial, and then covered three fingers high with the following water:

To 8 ounces of aquafort, 2 ounces of the antimony prepared as above, 1 ounce of the prepared sal ammoniac, and approximately 1 drachma of wine vinegar were added. When the antimony and the sal ammoniac had dissolved, then the aquafort was poured into a vial containing copper and allowed to dissolve for 3 days and nights. Then 30 grams of the black sulfur was poured into 120 grams of wine vinegar and 40 grams of aquafort, which had not formed a precipitate but which contained 30 grams of the sulfur and had a blood-red color. After the above liquid had been allowed to stand for 3 days and nights to complete the solution process, approximately 10 grams of this red solution were added, whereupon all of the dissolved material precipitated took place, with a pronounced gradation as is customary until the mark of copper had completely dissolved, which often requires up to 10 or 12 days without heating, since it dissolves so slowly that such gradation can easily be noted. After complete solution, finely-powdered brown lime was added, which had been washed with warm eater, dried thoroughly, and subsequently melted in a crucible, resulting in 1 mark of copper and up to 140 grams of good Hungarian gold, in all test samples of "Antimony of Saturn", as they might be designated. Everything which I have been able to make is, as always, to the Glory of God.

Another Process from Copper Pyrites as Described in Old Notes Which Beuther has Fortunately Located

To finely pulverized, even though impure, copper pyrites, washed and edulcorated, and placed in a copper vessel, was salt water, slowly heated to boiling, with vigorous stirring. An eighth of a can of unprecipitated aquafort was then added, along with a small amount of salt, and the mixture was allowed to stand approximately 12 hours until a sulfuric odor was detected. It was then neutralized with soda, after which the solution was decanted from the pyrites. Half of the decanted solution, while still warm, was poured into the iron, transforming the iron into copper. Other samples of copper were obtained from the precipitate produced by aquafort, but were not such fine specimen as those from the unprecipitated material. Taking 1 mark of the copper and fusing with 1 ounce of

Hungarian gold, and then pouring the product into an ingot mold, and adding 6 ounces of unprecipitated aquafort in a vial with 2 ounces of molten salt and allowing the product to cool by standing for 3 hours, and subsequently heating for 2 hours, so that the aquafort bubbles up with the salt, and is now ready to use, and so is allowed to cool and stand until another day. The aquafort was then decanted slowly from the salt, into another beaker, and subsequently 6 ounces of aquafort and 30 grams of yellow arsenic and 20 grams of alum (both pulverized) were slowly added and afterwards heated vigorously until the material dissolved. The product was then slowly decanted into another glass container, to which was then added 4 ounces of aquafort and 2 ounces of golden sulfur, with bubbling. The product was then divided into 3 parts, the first of which was poured over the copper, along with 1 mark of gold and allowed to dissolve 3 days and nights. It was then treated with 30 grams of 9 times distilled vinegar. One part of this vinegar solution containing the copper was allowed to stand for 3 days, whereupon a precipitate was noted. The other part containing the aquafort was subdivided into 3 portions, poured onto the copper until it dissolved, and allowed to stand day and night, until the fourth day, it was again poured into an equal amount of vinegar, until a precipitate formed. On the fifth day, 1 portion from the water solution in the glass container was added, dissolving the material in 3 hours, after which an equal amount of vinegar was added. On the sixth day, the third portion containing the aquafort, which was the only portion still remaining, was added and allowed to stand for 3 hours to dissolve it, after which the vinegar was added. On the seventh day, the third portion of the aquafort was added to an equal part of the material from the glass container and allowed to an equal part of the material from the glass container and allowed to become lukewarm before dissolving 20 grams of salt which had previously been added, and then 2 parts (2 drachms) of vinegar were added. On the eighth day, one part of water was added to the third portion, allowed to stand for 3 hours, and then poured into 2 volumes of vinegar. On the ninth day, the mixture was allowed to remain quiet until solution was completed; nothing was added thereto. On the tenth day, the last of the water was solution was added to the remainder of the third portion, up to 6 hours being required for complete solution, after which excess vinegar was added and then allowed to remain quiet day and night, making it the twelfth day. On the thirteenth, the water was gradually poured off, the precipitated portion being completely brown. The product was then edulcerated and dried, whereupon the material was converted into lead, or antimony. The transmuted copper was prepared in marked quantities from the copper pyrites, until 150grams of very fine and pure gold were prepared, without any gold having been added, for which blessing I give thanks to the Holy Trinity of God, the Father, Son, and Holy Ghost. As always, to God, be all Glory given.

Beuther's Jupiter Process

One pound of Jovis Puri (pure tin), cut in practicably small pieces, so that they could be introduced into a small flask, was added, along with approximately 1 pint of distilled vinegar, to the third portion of the aquafort, and a handful of salt, all of which were poured into the tin and allowed to stand for 5 hours, after which the clear liquid part was poured off, the tin remaining behind was thoroughly washed with water and again poured into a vial and covered with 3 fingers of aquafort. Then 20 grams of sal ammoniac were added, at which point a strong reaction began to take place. The reaction proceeds best when it is kept in contact with fire. Then 1 pound of activated mercury was added, the more slowly the better, using an efficient wooden stirrer, which works better in material which is as viscous as a pulp. After aquafort is again added thereto, a vigorous reaction sets in with bubbling up and the product is stirred further and again becomes quite thick. Aquafort is again added very slowly, until the mercury along with the tin has become a single body, like white chalk (somewhat due to the fact that some of the mercury remains undissolved). It was then allowed to stand until everything was dissolved. When both of the substances added have been completely dissolved, 2 ounces of iron are taken and covered 3 fingers high with aquafort, then the material commences to effervesce and a reaction takes place, until the solution becomes entirely red or brown and begins to foam. The red-brown liquid was then poured into a glass dish and allowed to set until the volatile liquid escapes and the reaction subsides and is no longer active. After again being allowed to stand for a brief period of time, it again commences to effervesce and then quiets down to the formation of a yellow color and a supernatant liquid above the iron, which is also blood-red. The process is continued by the daily addition of 2 pounds of prepared chalk and 40 grams of aquafort, and sufficient chalk daily to maintain a content of 10 grams of silver. In my opinion, such additions must take place about every other day, although for a total of 14 or 15 times, so that from every pound of metal, as tin or mercury, 280 grams of very high quality, stable, finely pulverized silver results. As always, to God alone be the Glory.

Moreover, at the very beginning 20 grams of the red material from iron and 20 grams of sal ammoniac can be added, which reacts at the same time with the tin and the mercury, however, no aquafort can be added until both substances are destroyed. When no mercury can be observed in the glass dish, then 20 grams of the red material from iron, 20 grams of sal ammoniac, and 0 (?) grams of aquafort were added every day, up to 14 days.

First Supplement or Appendix

The Philosophers' Stone

The author has carried out this work with his own hands and it is, according to him, so easy that whoever cannot duplicate this work cannot duplicate any other. Since it teaches what the mercury of the philosophers is and how it is obtained from the body and how the silver and gold are obtained from feces, by the use of fire, by fixing, and by amalgamating with mercury, and finally, in this way, arriving at the Philosophers' Stone.

Preparation of Fixed Luna and Sol

Chalk (lime) in a crucible, along with 3 parts of Calaminaris Stone, 2 parts of white arsenic (sublimed 2 o 3 times) 1 pound prepared salt, and 20 grams of the same white arsenic mixed with 40 grams of tartar (calcined until clear, white, and translucent), all materials thoroughly mixed together, were treated with finely powdered silver until the product was thick as a ducat and a hands' width, forming a layer upon the other material present, and initially one finger thick in the above-mentioned powder (after the addition of the silver). The powdered material was again added and the process was repeated, with adequate luting and cementing, using an appropriate flame. Nevertheless, the silver does not melt, even after 36 hours. After cooling, and subsequent neutralization of the product, the silver appeared to be black and as brittle as glass. The product was then thoroughly washed with water and allowed to dry. It was then placed in a small mortar and pulverized completely, carefully washed with ordinary water, and then allowed to settle, and the entire process was repeated until the supernatant water was completely clear. The water can be evaporated, so that none of the silver will be lost, even that which has been dissolved in the water. The resulting white powder, after treatment with borax (at which time it is in lamellar form) was cemented by heat and worked up as before, so that the silver is set free from the feces and the element is purified without decomposition, which can be tested by rectification of a half-ounce of the powder in aqua vitae (distilled brandy) or by dissolving in distilled vinegar, coagulated, and again dissolved, until no more fecal material is found. This silver should also be cemented by heating, in order that the elements become fixed, and that the silver becomes fixed, as well, that they will not be dissolved by aquafort (or become grey-colored when held at high heat) or have lost their timbre, but are precipitated by antimony.

After cementing, the gold also has a similar form, which alone is that taken (instead of arsenic) by sublimed red mercury, which cannot be struck or washed like the silver ca. It is not black at all, but by cementing the gold for so long a time, until no further loss in weight, then it is gold freed of all fecal matter and must be fixed by cementing 3 times more, so that all the elements are prepared.

Extraction of Philosophers' Mercury

First, from the silver, dissolved in aquafort, prepared from vitriol (sulfuric acid) and saltpeter, precipitated by salt water, neutralized with lime, dried, and then poured in a glass container, set on a tripod or in a furnace to calcine it or reverberate it to give it considerable heat, as when lead is to be melted. After standing 6 weeks, the corpus was opened, so the mercury could be separated from the earth and the oil, even when it remains with the earth.

The same can be done with the gold. The calx, by itself alone, must stand for 18 to 22 weeks in a reverberatory furnace. When the furnace is opened, there is so much gold that it amounts to 2 or 3 fingers above the bottom of the crucible, much like a mushroom. The mercury can be sublimed almost immediately, even with a gentle flame. Generally speaking, sal ammoniac can also be sublimed and this is also the case in the extraction of the salt.

Clarification of Mercury

Sublimed 2 or 3 times by vitriol or arsenic, the product appears as clear as crystal. When gold and silver have now been obtained, the product can be pulverized finely on porcelain plate with clear, transparent sal ammoniac, which is not damp, even when pulverized with 1 pound of prepared silver and 80 grams of sal ammoniac on a porcelain plate. Add vinegar, distilled 5 or 6 times from its phlegm, to another vessel and distill it from a bath over a calx, properly luted, to give 1 pound of calculated silver from 4 pounds of vinegar. After then allowing the product to cool before opening it, it was found that the vinegar, the sal ammoniac, and the silver were losing their odor, which was normally so strong as to be indistinguishable whether cold or hot. As soon as it was opened, the glass was removed and the contents were washed well and set aside for 6 weeks in a water bath, so hot that one could scarcely drink the water, because of the heat. After allowing the product to cool, the product was broken up into small pieces and distilled into a receiver by the use of vigorous heat, and subsequently set in ashes and heated gradually until the mercury sublimed with the sal ammoniac and was snow-white in color. After standing an additional 24 hours, prior to subliming, it was cooled, opened up, and the sublimed mercury was taken out and weighed so as to know how much sal ammoniac should be added thereto. The sublimed material was again placed in a glass vessel, sublimed still one more time, and then feces were added and the product was sublimed again until no feces remained behind. This is the body of material containing the oil or orpiment and was then taken out and weighed, so as to be able to better know how much mercury is present, how much calx that there had been, how much of the body of material and earth had been in the glass, and how much distilled vinegar remained to be dissolved in the clear water. After adding the feces, the clear supernatant liquid was poured off, coagulated, and dissolved again until feces no longer remained. At this point, it was coagulated again, and the salt of the earth was prepared and it was clear as a crystal. The sublimed mercury and the sal ammoniac were then finely pulverized with this salt in a mortar when dry and added to the glass vessel, and finally set upon a tripod or in a calcining oven for 6 weeks and heated, and then allowed to cool afterward in a cool place covered with a cloth. It dissolved to form a clear solution in 6 or 8 days. This is the Philosophers' Vinegar, said to be our mercury, but also given many other names.

By taking 3 parts of silver and 1 part of gold and cementing them in the fire, as stated at the beginning of the section on the Extraction of Mercury, followed by solution in aquafort, the calx residue edulcorated, and dried (to give sufficient material to treat with distilled Philosophers' Vinegar) set in an ash bath over a small fire (as when gold or silver was to be dissolved) the product was then mixed with 3 parts of calcined silver and 1 part of calcined gold. After pulverizing the above-named materials together and subsequently treating with 1 or 2 ounces of the aforementioned water and allowing to dissolve, 1 ounce of the blended calx was added and allowed to dissolve until no more could be dissolved, but remained undissolved as a residue. The water then contained its own nourishment and was the best solution that could be obtained, since it contained the dissolved mercury, along with all of the metals. This solution was then poured off into another glass container from the powder which still remained dissolved and was set in an ash bath and heated to the heat of the midday sun and even better (hotter), and covered with a double linen, whereupon it coagulated to give a grey powder. The latter was poured into a small glass vessel and covered. It was then luted thoroughly and allowed to stand until the grey color began to turn white. The heat was increased gradually to that of the sunshine of a June day and kept there until the material became as white as snow. On increasing the heat to that of the sun on an August day, spiegelgleisen-type crystals began to form, which appeared to be very much like glass. The heat was then held to such a point where such crystals no longer formed and the material was then said to be fixed.

No harm is done when sometimes no spiegeleisen-type crystals can be seen, since the material had been fixed by the heat of the small fire. By starting with stronger heating, peaks of a hand's breadth were noted, which was not good since it led to an increase in the amount of vapor in the glass and in the white fumes that escaped. Using a gentler fire, the material took on a yellow or red color and might indicate decomposition in the product, even though it is better to use a gentle fire than a much hotter one.

After being fixed, the white material is divided into 2 parts, one to remain white in color, the other to become red. The part to be red was poured into a glass vessel, which was put in the oven, and allowed to stand until what is to follow has taken place.

The material to remain white was also poured into a glass vessel, covered over with what will be later mentioned as "paradise water" (9 parts of water to one part of the material), properly mixed, and put in an oven until the stone (i.e., gem-stone) is completely white, which takes place before additional heat brings it to the red stage.

Paradise water is the active mercury extracted from the silver, which was prepared from the white material, a previously outlined, with the single change that pure tartar be used instead of sal ammoniac, which is translucent, but in weight it is comparable to sal ammoniac. The calcined silver accordingly poured into a glass or stone vessel and covered with distilled vinegar (distilled as previously explained). Then the mercury was added dropwise into an alembic. This affects the nature of the tartar. The mercury also reduces the amount of moisture present, as previously explained. From 9 parts of this water and 1 part of the fixed stone for whitening, a change was made, transferred to a flask with a still head attached, distilled, and the products washed. The other glass vessel, containing the part to become red, was set in an ash bath and heated by a little hotter fire (about equal to that of the sun in the middle of summer), such as one might use in the preparation of rose water, and this treatment was continued until it is heated sufficiently in its own broth and all of the paradise water had completely evaporated, leaving behind a dark grey powder as a residue. The fire was then made a little hotter until the powder began to turn white. By gradually increasing the fire the powder became completely white and sparkling like snow, so that when you see it you will become very happy and thank God for His gift that he has given to Nature.

Half of this white powder is put in a crucible and set over a fire, where it melts as easy as does wax. When completely melted, it is poured into a wooden container coated with goat's tallow to make it smaller, harder, and as translucent as a crystal. One part of this material, added to 200 parts of tin, was converted into the very best silver that the human eye has ever seen, thanks to God and the prayers of the poor.

Of the other half of the white powder, 1 part was taken and added to 100 parts of calcined silver, the whole washed and dried, and then pulverized, together with the well-dried powder. To 1 part of this powdered material and 9 parts of silver from the previously prepared paradise water in a hermetically sealed glass container, heat was applied as before. Since, at first, the paradise water was added only after the powder had absorbed all of the spirit it was continued as before until all of it was fixed and became white as snow, so that half as much again was taken for projects and the other part was subdivided and put into several glass vessels.

Rubification

Using the other half of the first fixed powder taken to produce the red material, the time required until the white material was obtained by allowing it to stand on the ash bath until the material became yellow (saffron) was 8 to 10 days, with the heat increasing a small amount only. When the powder does not darken in color it means that the fire is not hot enough, so the fire was increased a bit more for 10 days longer and gradually the product (powder) began to turn red. At this point, the heat should be maintained a little hotter for 5 or 6 days until the product becomes ruby-red and shines like fish eyes. The heat can then be made a little lower as the above sign continues, indicating that the product has become fixed. There is no hurry in increasing the heat at the very start, however, since the red color never comes before the yellow or the white color, but always ahead of the black, even though the effort might be in vain and must be begun anew, for it is certainly better to begin with a low fire than with a large one, since the regulation of the heat is half of this work.

When the material has become fixed and is now red as a ruby, then it is found to imbibe 9 parts of paradise water for each part of this material, just as occurs in every case with the white powder, although the paradise water from gold must be weighed out, just like its predecessor was from silver. With the paradise water, which must be luted well and then allowed to stand as before in an ash bath, and subjected to heat as when the lead was melted, without seeming (until all of the water had completely evaporated and a powder remained behind as a residue) to appear to be more black than grey, the heat was then kept the same until the product became perfectly white, and then the heat was increased only slightly until the color became a deep yellow and was continued at the same rate until no brighter yellow color could be produced, and then the heat was increased rather drastically until a red color appeared, until all of it became perfectly red, which was very fortunate, thanks be to God.

The half of this melt in the crucible was poured into a glass vessel and sealed with sheep tallow. It became very hard like glass and ruby-red in color. It was as transparent as a crystal. One part of this material, treated with 2000 parts of lead and subsequently melted, proved to be the best gold, of excellent color and stable in all tests, as well it might be considered to be the world over.

Multiplication

The other half of the powdered material also increased several times. Using the cemented gold, as noted above, it was dissolve, neutralized, and dried with lime (200 parts of lime to 1 part of stone), pulverized and then dried, then poured into a glass vessel and covered with 9 times as much as the pulverized material in the glass vessel. The material was then well luted and the heat regulated, as noted earlier, until the material we succeeded in multiplying the aforesaid amount of material and never ceased to be thankful for the prayers of the poor and to give thanks to God for His gift.

Production of Ruby-Red Water

Crocus Iron Antimony --- Mineral: Red arsenic, amount used, 1 pound orpiment, 1 pound Roman vitriol, and 3 pounds o saltpeter, mixed well together and then treated with a half-pound of sal ammoniac, and poured into an earthenware vessel, which can withstand a very hot fire, and also not become vitreous in the interior, was poured into a still, well washed, and distilled carefully, at first at low heat, then with gradually increased heat, until a white liquid distilled over. When such began, the color of the material in the still turned yellow or red. The receiver was then removed and replaced by a receiver with a better seal. The heat was increased until white spirits came over, which proved to be sal ammoniac, which was the first material to distill over. The heat was increased so long as the still head remained clear. When a red liquor started to distill across, but only in small amounts and then spirits and water distilled over and the distillation apparatus became white on the inside, as if full of snow. It was then carefully watched until the spirits and the water were drawn through the distillation apparatus with frothing, like as if a man was vigorously blowing his nose into the receptacle. When the spirits continued to distill and the distillation apparatus began to clear up, and the death's head (colcothar, red ferric oxide) began to form as a fine powder. On standing 2 hours in absolutely clear water and afterwards being brought to a boil, the clear water was decanted from the residue, which was then discarded. When the water was evaporated, a yellow material remained behind. The yellow powder was weighed and the calculated amount of saltpeter was added and the thoroughly mixed materials were put into a glass flask, covered with distilled water, warmed gently in an oven, and a distilling apparatus with receiver was set up. The distillation was carried out using gentle heat at first, then subsequently more vigorous heat, until a white spirit distilled over. A red water then distilled across, which was bright both by day and at night and which dissolved the fixed silver, and gave the most beautiful tint to the gold, and even gave to all of the white metals the color of gold. Praise God!

Other Work Done During a 12-Month Period on 100 Significant Tests, Whose Projection is Infinite

Four pounds of Roman Vitriol, which should be blue and clear, and saltpeter were mixed together and then dried. They were then evaporated together in an earthenware dish, using a wooden stirrer to start a reaction. Afterwards 1 pound of mercury, freed of excess moisture and using heated materials, were mixed well together and dried on a stone and warm strip of cloth, until no further moisture could be removed from the material, which was then poured into a flask to which a distilling apparatus was attached. After careful luting and allowing to dry further, a small fire was initially used and gradually increased over a 4 hour period. The water was distilled over in a period of 16 or 20 hours or more and afterwards the heat was adjusted and this resulted in a better process, although the same resulted in requiring 30 to 36 hours for the liquid to distill over, since by such a procedure the glass apparatus was much less likely to get broken. This is the common rule in distillations where aquafort is involved.

When the water has all been distilled over, the mercury has been found to be sublimed in the upper hand of the still, as white as snow and somewhat hard, and retains the tincture of vitriol with it, when not further sublimed. After cooling, the aquafort is put in a safe place and the mercury is removed from the distillation apparatus. It is pulverized further with the other half of the material and then distilled and sublimed as before. Again 4 pounds each of saltpeter and vitriol, proceeding as before with the mercury which had been sublimed four times and so was a complete tincture. Now 80 grams of silver, finely cupellated, 40 grams of cemented gold, and 60 grams of sublimed mercury (taken in the proper proportions), covered with 160 grams of silver, made by using aquafort as previously noted, were placed in a glass vessel, and 120 grams of the gold in another glass vessel, together with 10 grams of pulverized sal ammoniac, covered over with 60 grams of the mercury and the mixture of all three was set in a beaker on an ash bath to dissolve. Each flask has a cork stopper, so that air can reach it and, in addition all of them should be unstoppered every 2 or 3 hours, for the same reason.

Nutritional Material

Material in the subsequent form must be increased so that it is not destroyed by the aquafort. When 2 or 3 pieces of finely powdered, clean silver are put in a glass vessel and dissolved and the flask is stoppered and then sublimed mercury is dissolved cleanly with the gold, and more is gradually added until no more is dissolved after 10 to 12 days, then it is sufficiently nourished and has lost its corrosiveness. This is one of the great secrets of the art, since it must at all times be nurtured (sustained) on hot ashes.

When it is also nourished by pouring on each of the small pieces in the glass flask, then from each must 1 part be fixed and separated from the feces and filth.

Now, in the glass, along with the silver, there is a black powder due to the material remaining as residue and this is the gold, for in all dissolved silver is found powdered gold, since all silver contains gold, some more than others, to preserve such powder.

When you now pour each of them into a flask and attach to it a distillation apparatus and receiver and slowly distill it from a sand bath until dry,

then a good rectification results, giving 3 or 4
fingers of aqua vitae (brandy) lying supernatant
above the other material in the flask, then the
method is the best that can be set up using on all
occasions a distillation apparatus. However,
distillation through a nozzle is carried out more
slowly and fixes itself better, since on the
surface, with a glass stopper or small glass cover,
it is easier to cover, and this is necessary for
solution, coagulation, and fixation to occur. Since
each specimen is now in a covered flask and the
alcohol is poured over it and it is placed in a
bath, with the receiver at the end of the condenser,
and it has been well luted and dissolved over a
small fire, when solution is completed the feces
remain behind and the supernatant liquid is decanted
into 3 other glass vessels of the same type, taking
good care to pour another portion of alcohol 4
fingers high over the feces. Again, after allowing 3
or 4 days for solution to be complete, it will also
dissolve, the clear liquid will then be decanted
off, as above. Nevertheless, each flask is covered
and each sample of feces is put in a particular
glass vessel, and then put in a safe place until
more feces are obtained. The solution containing the
dissolved material is then poured into a flask, as
before, and put over a bath, and the aqua vitae is
evaporated therefrom until the flask is completely
dry, whereupon fresh alcohol is poured over it and
the product is dissolved as before, which requires 4
days. The dissolved liquid is decanted off, each
into its own special glass vessel and other aqua
vitae is poured over the feces and allowed to stand
another 3 or 4 days to dissolve them, and then the
clear supernatant liquid from each is decanted as
previously and the feces treated as before. This

process is repeated until no more feces remain any longer in the solution, in the case of all three of the samples. The feces samples that were set aside must now be calcined, and from each of them the feces must be extracted, each from its own sample, and at this point they are all added together.

Calcination of the Soil

In the 3 glass vessels containing the samples of feces, which should be completely dry, each is luted into a small glass container, which is covered, and then set each one into an ash bath subliming dish in a special oven. The gold must then be calcined a few times or more. Feces from silver require 30 hours, while feces from mercury require 18 hours. The fire is regulated so that each of the samples of feces is kept at the same temperature, that is to say, that from gold requires 6 days, that from silver 30 to 32 hours, that from mercury 18 to 20hours, and thereafter cooled, so that they are sublimed, dusted off with a feather, and then each sample of feces is covered over with ordinary water, put on an ash bath, and then allowed to boil for 1 hour. After cooling, the aqueous product was filtered or, in the alternative, the clear liquid was decanted, fresh water was added again, and subsequently the product was boiled for one hour more, and then the supernatant liquid was again decanted or filtered through felt. In this way the soil was separated from the fecal material, which was now discarded. The water decanted off was evaporated and as a residue in each of the 3 glass containers was found a white or grey salt; the soil is again dissolved in freshly boiled distilled water, the solution was boiled, and decanted from the feces and evaporated.

The resulting salt is still whiter, and dissolves readily as long as the salt has no fecal matter remaining behind it. This is the so-called Philosophers' Soil.

Taking now each sample soil, in order to obtain its rectified element, which has been rectified without its separation, as reported above, which is among those elements brought to the study of medicine, it is clear that, because of their subtility, their greatest external property is internal and their greatest internal property is external, and thus reversed, so that all their disorderliness has thus been exposed.

Coagulation

Samples of soil are put each into their own glass vessel for each element, as instructed above. Consequently, 3 glass vessels, each well luted, are put into an oven, each equipped with distillation apparatus and receiver, to distill at low heat the aqua vitae therefrom, until the material begins to become thick, although not completely dry, so that it remains properly in solution, and thus can be seen from the material around there, which is somewhat drier and thicker than it was heretofore, and so is allowed to cool, although not completely, but yet while it is a little warm, it was poured into a glass dish, although even when it is a little warm it is thicker than honey and is thus not very nice. Each dish is supplied with a glass cover so that no foul air can enter. The dish was then set aside for 4 days in a cold place, until crystals appeared and on these crystals some aqua vitae still clings. The liquid was then poured off from the same into another dish of the same type, which could also be covered. The crystals were divided into 3 equal portions, each in its own glass container, while the glass container having the decanted aqua vitae was set aside to evaporate until the product became so thick as that of the previous time, which required 1 or 2 days and was still not completely evaporated, although the alcoholic spirits had passed off even when it became as viscous as oil when allowed to cool further in a cold place for 4 days to allow crystals to form so that, when added to the above and evaporated further, until no aqua vitae remained on the crystals and the material was dry, crystallization was found to be complete. In this way the gold, silver, and mercury were prepared and purified.

Conjunction of the Fixing Operations

Using a fixing-glass containing the powdered silver and putting it, hermetically sealed, into an oven in an ash bath and heating it until holding the fingers in the ash bath would result in a severe burn, as soon as the glass becomes hot, the powdered silver melted within an hour and the temperature rose, resulting in 1000 small veins. After the temperature had dropped again, the material became thicker from day to day as the material became fixed. At the end of the cooling process the veins had become so large that many could not be located in such large numbers, and hardly 15 or 16 small veins could be noted, while the heat was increasing to the point where a finger could scarcely remain in the ashes long enough to say a Hail Mary. When all the small veins had decreased in number, the heat was so strong that a finger could not even be held in the ashes. When the heat was held constant for 24 hours and there was, in this time interval, no longer any increase or decrease in temperature and the material was completely fixed (which should require less than 40 days). Moreover, the body is completely prepared and a medicine (in reality, perfected gold) can be prepared (the same gold) when a concerted attempt is made. In this way, an elixir can be made from imperfect metals.

Moreover, the process for creating gold crystals should be investigated for it is a unique process and the oven is large enough that all 3 of the glass vessels and their contents can be investigated simultaneously and be given at the same time the same firing procedure. However, you should know that the mercury (as noted earlier) must be clean and free of all contaminating material before it is ready for examination and that it still contains the spirit, the vitriol, and the saltpeter, which it had taken up during the sublimation process, while the distillation soil and all of the others are not yet fixed and are things incorporated in it that are not simple but are compounded, on account of which the mercury must be heated until broken down into simple substances, and in the minerals 1 part should be fixed and 1 part not fixed. It should be understood that with respect to silver, when no more veins arise, it is enough and the silver is fixed (and the same for mercury), so that it is 1 part that is fixed and 1 part that is not fixed. The spirits of the vitriol and of the soil from the mercury contain a fixed part. Moreover, there is no way to increase the heat to the point where heat alone can produce the small stone, although from very intense heat 1 part may attain a sufficiently high temperature to be sublimed, still most parts remain as a residue and stay fixed. When the mercury in the glass vessel shows no change at such heat, as has already been noted in the case of silver, then that parties fixed and ready to be added to the body.

In like manner is the case of the gold crystals to be treated so that they will all be fixed, as was said for the silver, and so it has been found to be a perfect medicine, even more perfect than the silver, for it has the power to transmute all imperfect metals into gold.

When gold, silver and mercury are to be prepared, they must be taken together and a stone made therefrom. Pour the mercury over the silver in a glass vessel and, while the mixture is still warm, lute it well and set it again in the ash bath and therewith the spirit will be united with the body. Then allow it to cool, so that the body is ready to receive the soul or ferment. So, in the name of God, take this ferment or prepared gold and pour it while still warm, into the glass vessel, wherein are silver and mercury, and cover it well and allow it to stand for 7 days in the oven, using the same heat as before so that it will incorporate the 3 and then allow it to cool, but while it is still arm pour it into a glass ampoule and then seal it and set it in an athanor or tripod for 40 days. At the end of 40 days, a transparent crystalline stone, dark red like a ruby, can be obtained from the oven, thanks to God and the prayers of the poor. This is of infinite projected vale, on account of its subtility (refinement). The highest projected value is likewise attached to silver and mercury, and to the gold, which is the most beautiful of all in color that any man has ever seen.

Multiplication

To prepare the mercury, as previously explained, up to the crystalline stone stage, of which 100 parts were added to one part of this powdered stone, together in a fixing glass vessel, covered well, and set in an oven in ashes, heated so hot that a bare finger could not be held therein, and kept at that level of heat for 15 days, after which it was poured into a glass ampoule, hermetically sealed, and set for 40 days in a tripod oven. Heated to the point where the hand could be kept a while between the inner and outer oven, at the end of 40 days the ampoule was broken and the noble stone was taken out. And we thank God with all our hearts for His gift, this stone of great value, as projected from the first.

I.N.J. Thesaurus of the World

(1) Common salt and Roman vitriol, 2 pounds each, were carefully pulverized and the powder dried in a dessicating vessel with stirring and use of a very gentle flame, until all moisture was driven off. Taking 1 pound of mercury in a beaker and adding the other material with thorough stirring using a wooden stirrer until all of the mercury was in solution, then pouring it into a sublimation apparatus set in a sand bath, using initially a low flame until all of the moisture is driven off and the material is completely dry. After heating over a higher flame for 12 hours, when all of the mercury was sublimed, the product was allowed to cool overnight. The

sublimed material was as white as snow. The sublimate, however, should contain the spirit of the red vitriol, which is often referred to as "invisible sulfur".

(2) One pound of vitriol and the same amount of saltpeter gave an aquafort which was added to the finely pulverized, well-luted material in an alembic and subjected to a gentle fire on a sand bath, and distilled. A viscid phlegm began to appear in the distillation apparatus, and the amount of heat was increased and then kept steady until the still head began to turn red. Water distilled over first and then the receiving flask became covered with a waxy material, which dissolved in warm water in a vial. After 2 hours, despite the fact that solution was not yet complete, a seventh part of sal ammoniac, or a little more, was added until solution was complete (although it is better is no sal ammoniac need be added at this point, however). The solution was put into a distilling flask and a distillation apparatus was attached thereto and the flask was set in an ash bath and aquafort began to distill over. After all the water had distilled across, the heat was increased and vapors of mercury began to form on the side of the condenser and eventually gave a powder as white as snow (and was referred to as Quinta Essentia). By increasing the heat still further, the material distilled over still further, and on cooling the Quinta Essentia was removed, leaving behind only the feces. The Quinta Essentia was again dissolved in aquafort and then sublimed, and the Quinta Essentia was now completely prepared and ready to serve as a medicine for the stones and is now indestructible, volatile and unfixed, as well as white as snow.

(3) Taking the Quinta Essentia (Mercury) in a glass container, it was set in a Philosophers' Oven in a dry room to be digested or calcined over a low coal fire for 8 days or more, whereupon the Spirit of Mercury was heated until solution became complete.

(4) After calcinations of the spirit in a sealed container, placed in a water bath, an alembic was set up and distillation of the Mayenthan (also called Virgin Water, which is the finished product of the Stone from the last preparation and has a bitter taste and burns the eyes as does vinegar, and is volatile, unfixed, and incorporeal, and volatile and incorporeal on digestion. He who would separate the other three elements of mercury must again add feces and digest in a drying room and then distill the product over a strong fire, using a bath regulated t provide increasing heat. This resulted in a shining oil, which was set in an ash bath on a subliming fire of good heat, yielding a red oil and leaving behind a black residue of soil which was good for nothing.

(5) This Virgin's Milk was put in a glass vial and hermetically sealed and then placed on a movable shelf of iron in half of the oven, between the fire and the glass vessel containing the material, so that the Virgin's Milk would not be disturbed by the heat, and the glass container was then set in the Philosophers' Oven to digest. By common sense, the vessel had a similar fire on all sides, so that the Virgin's Milk would show no flow. It is of great importance to have a moderate amount of heat, held constant at the same amount, so that the material would not turn black too soon, but if blackening

does appear, be glad, since it is a sign of proper digestion commencing and should be maintained until all color disappears and the material begins to turn white. Increasing the heat further will have no effect, for soon all of the material will be completely white, the White Stone.

(6) When the vessel is taken from the oven, which is called a Philosophers' Egg, with the material which is now white, then 1 part is removed therefrom and finely pulverized and better than usual is added. The other part is again sealed up in such a vessel and softened by a strong fire until the effect is very thorough. The product is then divided and heating gives a red tincture of latent sulfur-vitriol and makes it have a shiny surface. With nothing added and nothing left behind, this Stone becomes red through the power of fire. Note that in 2 different glass vessels the results can be different; in one giving a white product, in the other a red product, even though using a single fire in all of the previously described methods, except that in the one case to obtain a white product the glass vessel must be removed from the fire, while in the other case to obtain a red product the material is allowed to remain on the fire, albeit with appropriate adjustments of the fire within its time period, as previously stated. Note also that a few philosophers work with Virgin's Milk and pay no attention to the other elements. Some of them also mix the element earth with the other elements. For example: 1 part of earth, 2 parts of fire, 4 parts of lead, and 8 parts of Virgin's Milk and the resulting stone obtained is stronger than the red one and weaker than the white one. Both processes

are carried out using the same type of containers, ovens, and fires.

(7) Learn now the applications of this medicine. In the first place, the white tincture, which have the power to change mercury, lead, copper, tin, and iron into actual silver, while the red can give the metal the tinges of silver and gold. Moreover, it is also true that when metals melt and give off fumes, then white or red medicines will be produced therefrom, as many as are desired, and strengthening the fire increases the amount of melting and after an appropriate time, decreasing the fire leads to many tinctures which were produced and tinctured very deeply and to metals which were brittle, friable, and malleable. Even so, this is still medicine and many sorts of metals can be poured even though their color and malleability falls off. This work has also proved to be very fruitful in the case of many areas of medicine. Lead and tin have been shown to flow readily, although copper, iron, and silver must first be heated to incandescence and powders used in medicine must be capable of fine dispersion when applied.

(8) These medicines may increasingly come into greater use because of their effectiveness, strength, and other good qualities and they are frequently dissolved in baths and again coagulated in dry heat, so that frequently they can be very useful as coloring agents.

The Oven is also Prepared

A good type of adhesive material, horse manure, paper ashes, and hair clippings, treated with salt water and vinegar were made into a paste which can be fashioned into an oven, large or small, with a suitable door, the height of the opening of which was a span, or at least a nose, and there were 4 other apertures, one on each side, by which the smoke, fumes, and vapors can pass out, without collecting at the top of the oven. Thereafter, a thin metal plate made of iron was permitted to serve as a sliding door. Furthermore, there are 4 indentations for arms and pulleys to form a cross, set at the top of the oven, so that all of the heat will best be utilized. On the metal plate there is a glass vessel for preparing the adhesive material and a Philosophers' Egg to aid in the digestion process and in the heating of the Stone. The 4 sides of the plate were also carefully fitted into the oven, so that the rings can be about 5, 4, or 3 fingers in width, depending upon the width of the iron, but there will be at least one width between the side of the plate and the oven. A small removable cover can also be made of porcelain, copper, or iron. One can also use a small box, whose cover is removable, and mark the place where the material in the oven will be inspected. The cover will be well constructed, both inside and outside, to hold the heat and make it a suitable apparatus having a door. In this way a hand can be used to turn up or down the heat and so control it and the extent of the fire. At the beginning of the digestion period or of the heating of the Stone, such heat is sufficient if you may hold a hand between the containing vessel and the side of the oven for a good while. Take care that you do not get the vessels mixed up and so have to discard the flask in your analysis. You must cover it well at all the openings so that too much heat does not escape.

Preparation of Salt of the Earth or Saltpeter of the Philosophers from Virgin Soil

In the month of May, when the sky is clear and bright and the air is pure and the weather is still fine, without wind and rain, and lovely odors are everywhere, that is, as it were, lovely odors are rising all around, one should enjoy the early morning and the sunrise, along a beautiful path where the good black earth is so rich, and even more beautiful is the red, the gold, and the red-gold of Nature's beautiful plants and flowers, which take care to bear in the clover family. The red and gold earth in the mountains of the wine country is also very magnificent. Also even the lime-pit, when you can regard it as more than just sandy. Dig down a few hundredweights, shut yourself apart from one another, so that the stars might well operate therein: This, one should also let lie fallow for 14 days and nights; however, should rain set in, one must cover it well with straw until the weather becomes clear. After waiting 14 days, put the earth in tubs and cover it well. Afterwards, a few tubs should be made, like the silk refiners have, for subjecting the soil to washings with lye (warm water is best) and allow it to stand for 24 hours and then afterwards draw it off, boil it down tone-fourth its original volume, let it stand for a few days, and then open it. It is like saltpeter; it burns, but not much. It is then dissolved, filtered, and coagulated with rain water until complete. Hereafter, it is called Sal Terra (Salt of the Earth) and also Saltpeter Philosophorum (Philosophers' Saltpeter) or even Sal Natureae Virgineum (Pure Natural Salt) which is dissolved in the seas of the world. The hand is not easily washed.

This is now the high Secretum Philosophorum (Philosophers' Secret) in which the Universal Spirit of the World is often hidden. It is often called Woman's Work, since it is associated with soapmaking and cooking.

Sal hoc naturae (Salt of this nature) is thus a unique thing to be considered, since it is actually made up of three different kinds of salt. First, there is a universal salt of the nature of saltpeter and soil, in which the spirit of the world frequently resides. It is not volatile and not completely fixed, but has a middle nature. More than any other, it contains a sal ammoniac, which is actually volatile. And, in the third place, it contains hidden in it an alkali, and a fixed salt. It is also triune, and manifests itself in its subsequent reactions.

Preparation of the Universal Spirit and Volatile Salt; Sal Ammoniac and Salt of the Earth

Sal Superius (Superior Salt), 1 pound, is added to 3 pounds of special earth, from which the salt was made, although such earth must first be further calcined in a potter's kiln, then made into droplets with rain water (much like child's play). The droplets were allowed to dry and then distilled in a well-mounted retort above a controlled, but open, flame, by way of a condenser into a large receiver, wherein some pure water had been poured. The distillation was carried out using a hot fire. After the volatile salts had sublimed in the receiver, the part of the product remaining in the neck of the retort was gradually allowed to cool, the spirits and volatile salts present in the receiver were then washed together and combined and the combined washings were carefully neutralized. Again taking 1 pound of salt, proceed as before and repeat this procedure 4, 5 or 6 times, neutralizing the spirits and volatile salts carefully so as to incur no appreciable loss. The spirits and volatile salts were neutralized together in a large glass flask and separated from the phlegm on a steam bath and distilled over from a sand bath six successive times to give the spirits, which were always carefully preserved.

However, the salts remaining behind at the bottom of the flask after distillation over a hot fire was neutralized along with such as were sublimed. The process was repeated 5 or 6 times, and the product was white as snow, with no feces remaining behind, and was stored safely.

Now all Capita Mortua (dead heads) were taken and calcined one more time in a potter's kiln and the resulting fixed salts were extracted with distilled rain water therefrom and the best of such was purified until it was perfectly white, sparkling, and clear as a crystal, with no fecal matter left in the residue.

Composition of the Triune Principle

To have body, soul, and spirit separated from one another was brought about through the preliminary work involving fire, and also from heterogeneous, liberates mankind from the curse of the earth. Now we will examine such things together.

Volatile and fixed salts have so much in common. They can be pulverized in a glass mortar to a very fine powder. Put in a glass flask, with spirits of our salts added separately and then combined all together and well mixed, were subjected to digestion by the gentle heat of an ash bath for 8 days and nights and the three such principles were united rather than pleasantly again with one another in a single solution. Some feces were added, and then subsequently separated out, giving the proper master key of the philosophers in hand, with which all hidden metals, minerals, and precious stones are unlocked and brought to their quintessence. This then is the true universal measure, the proper water of life, with which all metals can become readily combined, especially the gold, which can be made alive, and the fountain, in which all metals are easily bathed and again become alive. It is the bath of their regeneration, their proper original material, and the most harmonious, most beloved mother, from which they all originated in the first place, and they again enter into life with great joy and pleasure, entering into the new and better life into which they have been reborn. It is the great salty sea of the world, the water bridge which, from the earth, serves as a general bridge over which everyman must cross, whoever was created. And thus, it is with the spoken word. It is St Basil's All in All, for in it everything extends and everything arises from it. It is the true being of all things. It is our Philosophers' Virgin Milk.

Subsequent Work and Composition

When gold is desired, then it is filed 7 or 8 times from fused antimony, the finer, the better (1 part) and 10 parts of Aqua Mercurii (mercurial water), mixed together together in a separating flask, the flask is then sealed, and set for a day, and a night on an ash bath, whereupon the gold is completely and readily dissolved. The earth, situated at the bottom, is separated therefrom. The clear supernatant liquid is poured into a vial, sealed hermetically, and set in the athanor oven. At first, Ignis Vaporis (Fire Vapor) was noted, until is became black, which happened in the space of 40 days, and then in the ashes a somewhat stronger fire was noted, so that the glass became so warm, like the sun at midday in summer shines on the surface of the cover, so that in the following 50 days all kinds of beautiful colors can be seen. Afterwards, set in a sand bath, as deep as a dollar's thickness, it remained only broad enough to show subsequently after 50 days the bright light of the diadem (the moon). Therefore, do not cease, but be sure to continue your Gradum Ignis (degree of fire), for in this way all of the material will be yellow in 30 days. At the end of that time the glass vessel should be entirely buried in the sand and when heated strongly, the product will become blood-red in 40 days and in the middle it will appear as a luminous ruby, so Nature has distinguished them apart. It is seen here that it will not become a well-known tincture, but it distinguishes Nature from the heterogeneous. This is the same that one sets aside when it is of no use (although it can be made use of in some chemical analysis and separations.

The Composition of Precious Stones with Their Own Liquids

Detection of Mercury --- The salt of the earth is rather magnetic, that is, it shows the spirit of the air. According to the preparation of our water or Water of Neptune, in which the gold melts like ice in warm water, then it is the proper key, which gold opens to take its soul away, as a well-known Sal Ignis (Salt of Fire) of Nature, an imaginary spirit, the dragon, greedily devouring its own tail, the non-combustible sulfur. Note that when one makes mercury in the spirit of the world, then he is toying with time in so doing, and has hereafter an entry in gold, then the gold itself is also the Spirit of World red mercury coagulates.

A bath was prepared using our Sal Aquae or Spiritus Mundi increased to 9 degrees in mercury, such as that which was indicated by Philalethes by 9 eagles, representing that all metals were purified to the highest degree and are, apparently like pearls. Fortunate indeed is he who is able to conclude a bargain.

Our material in the atmosphere around is the mercury of the philosophers and the true mineral and vitriol and we have deducted a few hours to active metals, helping them to combine them in a flask and to putrefy them by the use of gentle heat, to which purpose a white clay is added at the end, which improves the quality of the precious stone. By redistilling on a steam bath, the material can also be converted in a few days into an active metal, using the above and carrying it out in a flask. Our

material includes everything, earth, air, and water,
all combined in one.

Second Supplement or Appendix

Clarification --- A Very Useful Illustration

Alchemical Art --- The Extent of its Great Benefits
During a One-Year Interval. Comparison of Costs of
Publication. Abstract of Profit Ledger. Practical
Approach to the Problem of Processing Food. Which of
the More than One Thousand Processes so German and
so Clever Have Not Been Described.

An excellent and very useful and true examination of
the art, using a few grams of fine gold and other
less important conditions and the lighter work load
involved as a result of the properly conducted
distillation techniques in a very short time.
Approximately 6 or 7 weeks are required to make a
few grams of the tincture and to prepare it for use,
as actually are clear in what is to follow.

Using 10 grams of this tincture to compare 1 mark of
fine silver to truly natural good gold, as God has
created the Earth and can add color to the same, and
this in all tests remains proper and authentic,
since aquafort cannot attack it or cause damage to
it or to the antimony glass or is itself damaged in
any test. The same can also be carried out better
with such a tincture, so that when 100 grams of
mercury, purified as before, were added to 10 grams
of the same well-purified gold in the same form as
previously noted, all of the test samples easily
passed all of the tests that were made.

The same work has been carried out at least 7 times
in a single year and each time 110 grams of the
reported tincture were made, so that in a single

year the total was 770 grams, of which 10 grams were used to make 1 mark of silver into gold, so that in a year 77 marks of gold at 160 per mark. At 10 grams for 8 gulden and each mark estimated at 128 guilders, totaled 77 marks of pure silver, the process can earn 9856 guilders.

From this sum of money, at a cost of 56 guilders for 70 grams of pure gold (at 8 guilders per 10 grams), more than 739 guilders and at 3 batzen was the estimate, and 8more guilders for the glass apparatus, fuel costs, and other expenses, so one must make in a year by this process, and subtracting 2 guilders per week for the wages of the laboratory workers and 1 guilder for the cost of the laboratory bench makes 130 guilders, making the entire cost of the year 933 guilders and 3 batzen. Subtracting that amount, there is still a profit of 892 guilders and 12 batzen left, after deducting all expenses, so that one man will have that amount per year, by what this endeavor lays out and can accomplish.

And when one man, using the tincture, might by the following process, being allowed to work with the mercury, it would cost him no more than 6 batzen, in addition to the 77 marks, 30 guilders, and 12 batzen. So this man has a projected overall income for the year of approximately 9631 guilders and 3 batzen.

This magnificent, grand, useful, and exalted artistic endeavor had a man from Venice, who also related to me that in this process and the work in carrying it out as follows hereafter, to work in his laboratory, and be allowed to carry out the tests which were deemed appropriate. And such artistic

endeavor was also taken to England, where the King
of England himself also tested it and found it to be
just exactly as claimed, thereupon approximately
20,000 guilders in gold was set as the sale price,
and subsequently testing of this artistic production
was carried out in the laboratories of Venice and
England, by Polish and Bohemian workers, and by the
same workers in the laboratories of Danzig, Prussia,
each separate one costing about 2,000 ducats,
without all of the tests being done, since there was
already good information about these tests.

Now follows the information about the process,
heretofore passed over, referred to as the Art of
Preparing Tinctures. In the Name of Almighty God:

To start with, using 3 pounds of white Salt from
Halle in a Waldenburger flask and pouring over it 1
measure of distilled vinegar then sealing the cover
of the flask with prudent luting prepared from very
adhesive glue, and then allowing the flask
containing the salt to sit on a warm sand bath or
coal stove until the salt becomes completely dry,
when it is then ready. Note that this is the salt
and vinegar, as hereafter follows, that now brings
about the mixing of the mercury with the salt, as
will hereafter be reported.

Now followed the method whereby the vinegar should
be distilled: Taking 2 measures of good vinegar, to
which are added 2 handfuls of ordinary salt, and
putting into a distilling flask such as aquafort is
distilled in, and continuing diligently, so that no
black material distills over (and if such should
take place, remove the receiver immediately), for
the material remaining inside the distillation flask

should still distill over as a clear distillate. Now ass another handful of salt to the clear liquid and continue the distillation as formerly. After 3 such distillations, the vinegar is properly prepared, the salt is collected, as noted above, and you will now see the Quintessence of all metals and the mercury of the philosophers, which is found in all places and is by no means by the mercury which is here referred to as living, which is toxic, but which is found as a salt to be something that no man can do without.

In the Name of God: Now take again 3 pounds of good dry white salt of Halle and add thereto 60 grams of the above steeped salt from the flask, mix the ingredients well, then put them into a sturdy glass retort or flask, glazed on the inside, and put the flask into the oven on the well-preserved side. Then attach a glass receiver, after putting in the receiver half a can of distilled water, well-covered, so that the glass receiver will be attached to the retort or the distilling apparatus.

After heating over a gentle fire so that the flask will not break or even begin to be damaged, since such a gentle fire needs an entire day, so for the next few days, the fire should be somewhat hotter and the flask will glow with a dull brown color rather than with a bright color, and the salt doesn't melt, but only appears as white fumes. This is the so-called Quintessence of all metals, whose ores are put into water and allowed to remain at glowing heat for 8 days, until a white material begins to sublime at the top of the receiver. This is the Mercury of the Philosophers. After distilling 8 days, the fire was removed and the white material was rinsed from the neck of the flask into the water and the rinse water was poured into a glass flask. A still head was attached to the flask and the material was distilled over an ash bath and in 12 hours the water had distilled over and tastes sweet, although it soon becomes sour like vinegar, becomes then stronger with time gradually, and finally sets the teeth on edge. When the distillation is finished, the residue is removed from the fire and allowed to cool. This is then the Oleum Salis (salty oil).

Note, too, that the sweet water distilled over in half a day and on cooling crystals formed rapidly in the water. The crystals were long and pointed and much like asbestos, but fine and sweet. This is the Mercury of the Philosophers.

The water from which the crystals were taken was distilled again for half a day, as before, and then allowed to cool.

Again crystals formed, as noted above. When no more crystals were formed, the water was again distilled off and the salty oil was obtained, as before. You should be aware that this Oleum Salis is the gold and when finely dispersed and dissolved that it is the water contained in all metals. It is also able to extract the sulfur from all metals and to serve as a medicine for men and metals.

The tincture is made as follows: 70 grams of the above Oleum Salis and 10 grams of finely pulverized gold give a yellow solution in water and yield 30 grams of the sweet crystals. The glass container is carefully sealed and set in a warm place until coagulation takes place and a hand material is formed. This hard material is then finely pulverized and put in a glass container which is then set in a steam bath, until the soft oil which is formed soon hardens. Distillation of this product from a steam bath until no more fumes come off and the material flows like wax, indicates that the product is finished. This amounts to 10grams, which make up to 1 mark of fine silver. While still hot, and allowing to flow together for a good half hour and then pouring into a tin container, the result is genuine natural gold, which God created and put in the earth, tested and true, which aquafort can never attack and which never does damage to antimony glass, and is never damaged by experimentation of any sort.

Note also that the Waldenburger flask must be well coated with a good luting material and subsequently carefully dried, otherwise it will break apart, so that the flask will not be preserved and the vapor will secretly be lost and the Oleum Salis will be greatly weakened, so that the Quintessence of the live mercury will be drawn of and lost. This is prevented as follows: Taking 1 pound of sublimed mercury, finely powdered and put into a well-luted flass flask, covered over with the above Oleum Slais and stoppered well, and heated 8 days, until the mercury is separated by distilling from the Oleum Salis. By heating very strongly the mercury, the residue in the glass becomes melted. After the fire is removed and the flask cooled, crystals form on the walls of the container. This is the Quintessence.

The very fragile gold is also extracted from the tincture as follows: Taking 70 grams of Quintessence, made from the clear crystals, as noted above, and dissolving it in 40 grams of Oleum Salis, and adding a solution of 10 grams of finely pulverized gold dissolved in 70 grams of Oleum Salis, gave a product which was then treated with 10 grams of sal ammoniac and allowed to stand until all of it dissolved. The resulting solution was poured into a well-luted glass-stoppered flask and a distillation apparatus was attached to the flask. The material was distilled until dry and the dry residue was sublimed, at which time the Quintessence appears as red, and then blue, material. After cooling, the proceeding is identified as gold, the Quintessence. The dry product was allowed to stand on a glass plate in a warm spot until a red oil flows out. When the oil hardens, as was the case with earlier tinctures, then one part of this material was added to 10 parts of mercury, and when heated until fumes form, it becomes hard and is pure gold, which passes all tests for the same, for which God will eternally be praised and thanked, as He was also in the case of all other metals.

The adhesive material used to coat the retort and the glass flask: Take the viscous adhesive material, which is neither sandy or gritty, moisten it with the water in which it had earlier been paced, add a handful of finely pulverized tartar, mixed with hair, horse manure, and common slat, until the adhesive material is very sticky. Pay careful attention that no sand or other impurities are included; store it under a retort or other glass vessel to become thoroughly dry before use or before setting near a fire.

Mystery of Mysteries

Here begins one of the great mysteries of Nature,
wherein are to be found great works of wonder. So
say that philosophers and also the wise people of
the world as well, about these Seeds: "Do you not
know when you work in this way with metals that a
heavenly influence according to one of God's
ordinances has been spoken to man and blends with
earthly characteristics? When, now, the two happen
in conjunction with each other, then an earthly
pattern is formed, which underlies the origin and
seeds of all metals". When such a lack of
understanding shall be stated that in the air there
is a slat from which all creatures originate and
grow and can be supported, without which nothing can
take place and that one can, by the heat of the
sunshine on the open field, one can obtain a
beautiful white salt in all rooms and living
quarters of his house and, furthermore, in salt free
water in appropriate amounts, with which wonders can
be accomplished, without doubt one can laugh
defiantly and say about the whole matter: "This is
very foolish talk, for nothing is accomplished by it
and either of us would be a fool, since it is
contrary to the truth when one wishes to proceed
thereby":

Take a perfectly round reflecting mirror, a weighing
scale of proper size and depth so that the rays of
the sun can be focused on a spot in the middle, then
set up this mirror when the sun is giving of its
hottest rays, under a cloudless sky; and allow it to
stand several days, removing it at night, and
protected from the wind and rain, and then a
beautiful white salt forms in the middle of the

mirror, which is shaved off by means of a knife in a clean glass and collected, in a short period of time it can amount to several ounces and then increase both in appearance and weight, and in this salt all 4 elements are found, from the air it has become conceivable when set in the sun's heat that it flows like water, the fire brings out the taste, and when allowed to stand for a long period of time in decaying matter, it turned blood-red like a fiery red ruby.

The water is allowed to stand in the hot sunshine in an open field or at home is a glass with an average fire under it, and then set in the hot sand, so the heat will draw the air into itself, and added to the coldwater in the glass. This water was then increased to a good amount and serves the Philosopher as his vinegar or solvent. This water also contains the 4 elements. From air it is turned into water. On concentrating by boiling, it turns into a fiery-red ruby. This water can be made to assume either a soft or hard form at will. And again it can serve as a very fine medicine for all sorts of illnesses. So, it is prepared for itself alone, or it can become the great Elixir of Nature (the secret remedy), which God Himself has ordained. Made by putrefaction and distillation, it will thus take effect more quickly and with it you can accomplish great wonders, as suits your pleasure.

The Testing of This Salt

Take half an ounce of this salt and add 3 parts of water. Then pour it into a vial and set it in very warm water, where the salt will melt as ice does in water. Allow it to evaporate to about half of its original volume. Put the remaining volume in a clean glass flask and set it in a cold place. Then separate the completely white, transparent crystals which form, which melt as soon as they are allowed to warm up. The crystals are beautiful and sweet, and are placed in a silver dish and set on a sand bath. The gold leaf or silver leaf will now dissolve, so that one does not see anything remaining. This is the real Aurum Potabile (potable gold) and it can be used I very small dosages for all types of sickness, and even more, it will take care of everything better than a person would believe possible.

Now For Particular Effects

Now take the gold, purified by the use of antimony and beat it as thin as possible or let it be put through a gold-beater to obtain as thin a gold leaf as possible, or as desired, and then add the crystallizing solution and then pulverize as much of it as desired. Then set it in a warm ash bath, whereupon the gold radical will dissolve, as will the silver. When the gold or silver has dissolved in the crystallizing water, then allow the solution to set in a dark place for 3 days. The glass flask must be well luted. Coagulation of a salt will then take pace in the flask and afterwards the solidified salt will again be added to water, as much as is deemed adequate. The gold is now in solution and will be blood-red, while the silver will be sky-blue. Then 1/2 pound or one pound of mercury is taken, or in any desired amount, after which there is much of the noble water. It is then poured onto the mercury, so that the latter is covered over to a depth of 2 or 3 fingers. As a result of this, the mercury will dissolve in the water. After solution, in which the gold is also dissolved, the mercury will be precipitated by a few more drops, added one after the other, and so will settle on the bottom. After decanting off the aqueous layer, then take up the mercury remaining as a residue and then drive it off from the lead, and as a result you will have the highest yield of gold that you can have, while in the ferment of silver, you have the most finely divided silver that can be seen by the naked eye. This is the uncontradicted truth that, as a matter of fact, even in this aqueous solution of gold, the copper, tin, and lead can be transmuted into gold, while the ferment of silver yields only silver.

Now to the Principal Work

Now take the aqueous solution of gold and the gold after 7 purifications and gradually add 3 times the amount of water necessary to fix the gold. Then place the well-luted flask in a bath or even in a kettle of warm water for the period of one month. It will then be as black as black velvet. Then take the flask from the water and set it in a warm ash bath. Then it will become white and fixed in a month. Then allow it to stand for 14 days after which the resulting stone will have a bright red color in a strong fire, like a ruby, and will be transparent. Moreover, if you have God's blessing, you will be able to accomplish by this technique a work which scarcely one person in 10,000 will believe.

Note well, this great work shall be kept secret to the end by the pledged word, so that it does not fall into unworthy hands, even if it requires overlooking a lot of misfortune and distress.

Your everlastng enlargement is also: One dram poured
out of its own water, that is to say, water of
crystallization, 2 parts, allowed to stand 8 days in
lukewarm water, then in hot water for an additional
8 days, then in even hotter water for a third stage,
and even a fourth stage, until a stone is formed.
The more often the heat is increased, the nobler the
stone will be. In ths way, the process can be
increased forever, for which yoou should love God
and also help the poor.

Note well, that the process described here, if
allowed to fall into unworthy hands, will lead to
nothing important at all.

Finis.

A Word from the Publisher

Thank you for purchasing this small work from The R.A.M.S. Library of Alchemy. During his lifetime, Hans Nintzel was dedicated to the identification, acquisition, study, retyping and, when necessary, translation of what he considered to be the most important known works on Alchemy. Hans was assisted by his sparse network of fellow Alchemists, all members of the Restorers of Alchemical Manuscripts Society (R.A.M.S.). I was an active member of R.A.M.S.

My goal is to publish all of the works originally made available through R.A.M.S. as photocopies. To facilitate this, I have chosen to have the books professionally printed. I also have a few titles that I intend to add to the original R.A.M.S. Library, selected by strict criteria established by Hans.

The works from the original R.A.M.S. Library are republished by R.A.M.S. Publishing Company in the collection, "The R.A.M.S. Library of Alchemy," with permission of the Estate of Hans W. Nintzel.

If you have a work on Alchemy that you believe should be a part of the R.A.M.S. Library, please contact me through R.A.M.S. Publishing Company.

Philip N. Wheeler

www.ingramcontent.com/pod-product-compliance
Lightning Source LLC
Chambersburg PA
CBHW080810180526
45168CB00006B/2390